DK HANDBOOKS

DINOSAURS

AND OTHER PREHISTORIC LIFE

DK HANDBOOKS

DINOSAURS

AND OTHER PREHISTORIC LIFE

HAZEL RICHARDSON

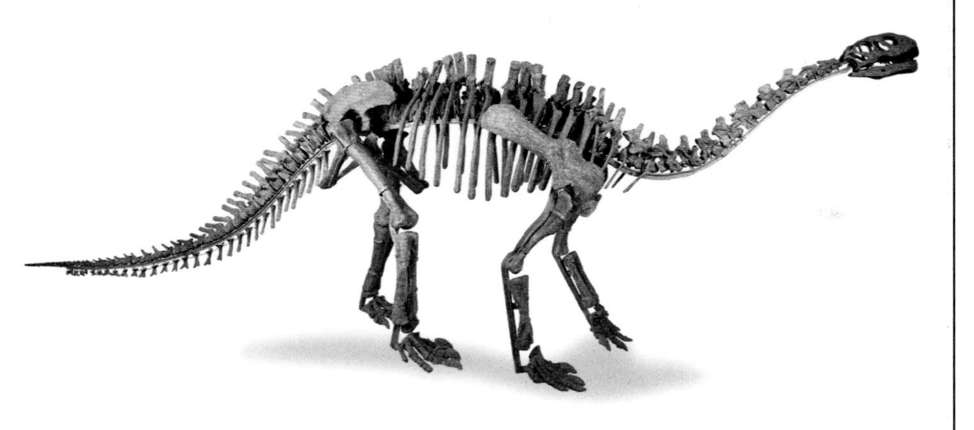

Consultant
DR. GREGORY F. FUNSTON

NEW EDITION

DK DELHI

Senior Editor Suefa Lee
Senior Art Editor Mahua Sharma
Assistant Editor Nayan Keshan
Assistant Art Editors George Thomas, Aarushi Dhawan
DTP Designers Umesh Singh Rawat, Vijay Kandwal
Senior Managing Editor Rohan Sinha
Managing Art Editor Sudakshina Basu
Production Manager Pankaj Sharma
Pre-production Manager Sunil Sharma

DK LONDON

Editor Annie Moss
Managing Editor Angeles Gavira Guerrero
Managing Art Editor Michael Duffy
Jacket Design Development Manager Sophia MTT
Senior Production Controller Meskerem Berhane
Production Editor George Nimmo
Associate Publishing Director Liz Wheeler
Publishing Director Jonathan Metcalf
Art Director Karen Self

Consultant Dr. Gregory F. Funston

FIRST EDITION

Project Editor Cathy Meeus
Picture Researcher Mariana Sonnenberg
DTP Designer Rajan Shah
Senior Editor Angeles Gavira
Managing Editor Liz Wheeler

Art Editor Lee Riches
Designers Tracy Miles, Rebecca Milner
Production Controller Melanie Dowland
Senior Art Editor Ina Stradins
Managing Art Editor Phil Ormerod

Consultant Dr. David Norman

2/3D Digital illustrations Jon Hughes
Additional 3D content Russell Gooday

This edition published in 2021
First published in Great Britain in 2003 by
Dorling Kindersley Limited
DK, One Embassy Gardens, 8 Viaduct Gardens, London, SW11 7BW

10 9 8 7 6 5 4 3 2 1
001–322044–Oct/2021

A CIP catalogue record for this book is available from the British Library.
ISBN 978-0-2414-7099-2

Printed and bound in China

For the curious
www.dk.com

This book was made with Forest Stewardship Council ™ certified paper – one small step in DK's commitment to a sustainable future. For more information go to www.dk.com/our-green-pledge

Contents

HOW THIS BOOK WORKS

This book opens with an introductory section that describes the evolution of early animals, followed by a summary of the environment on Earth during its early history. The main part of the book contains profiles of the major prehistoric animals from the Mesozoic and Cenozoic Eras.

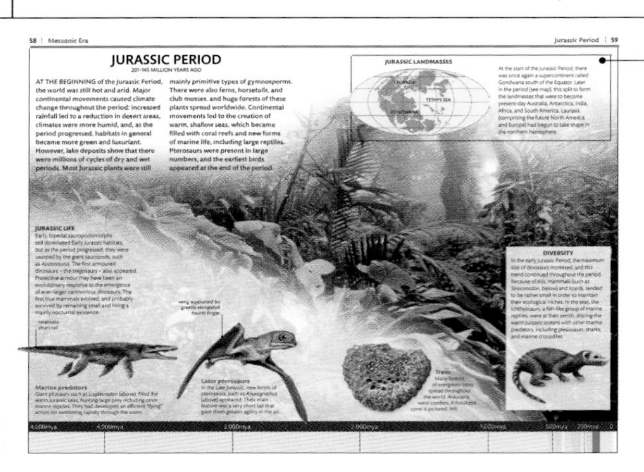

58 | Mesozoic Era

Jurassic Period | 59

JURASSIC PERIOD

201–145 MILLION YEARS AGO

AT THE BEGINNING of the Jurassic Period, the world was still hot and arid. Major continental movements caused climate change throughout the period: increased rainfall led to a reduction in desert areas, climates were more humid, and, as the period progressed, habitats in general became more green and luxuriant. However, lake deposits show that there were millions of cycles of dry and wet periods. Most Jurassic plants were still mainly primitive types of gymnosperms. There were also ferns, horsetails, and club mosses, and huge forests of these plants spread worldwide. Continental movements led to the creation of warm, shallow seas, which became filled with coral reefs and new forms of marine life, including large reptiles. Pterosaurs were present in large numbers, and the earliest birds appeared at the end of the period.

JURASSIC LANDMASSES

At the start of the Jurassic Period, there was once again a supercontinent called Gondwana south of the Equator. Later in the period (see map), this split to form the landmasses that were to become present-day Australia, Antarctica, India, Africa, and South America. Laurasia (comprising the future North America and Europe) had begun to take shape in the northern hemisphere.

JURASSIC LIFE

Early, bipedal sauropodomorphs still dominated Early Jurassic habitats, but as the period progressed, they were usurped by the giant sauropods, such as *Apatosaurus*. The first armoured dinosaurs – the stegosaurs – also appeared. Protective armour may have been an evolutionary response to the emergence of ever-larger carnivorous dinosaurs. The first true mammals evolved, and probably survived by remaining small and living a mainly nocturnal existence.

DIVERSITY

In the early Jurassic Period, the maximum size of dinosaurs increased, and this trend continued throughout the period. Because of this, mammals (such as *Sinoconodon*, below) and lizards, tended to be rather small in order to maintain their ecological niches. In the seas, the ichthyosaurs, a fish-like group of marine reptiles, were at their zenith, sharing the warm Jurassic oceans with other marine predators, including plesiosaurs, sharks, and marine crocodiles.

Marine predators

Giant pliosaurs such as *Liopleurodon* (above), filled the warm Jurassic seas, hunting large prey including other marine reptiles. They had developed an efficient "flying" action for swimming rapidly through the water.

Later pterosaurs

In the Late Jurassic, new forms of pterosaurs, such as *Anurognathus* (above) appeared. Their main feature was a very short tail that gave them greater agility in the air.

Trees

Many forests of evergreen types spread throughout the world. *Araucaria* were conifers. A fossilized cone is pictured, left.

a map and description of the positions of the continents during each time period

introduction to prehistoric life for each time period

THEROPODS

…transition …osystems, …seudosuchians …the early …ds, became …e ceratosaurs …ated the food …ge spawned the … top predators …e until the …. Tetanurans …th some groups …ts and horns, and others an elongated snout – taken to an extreme in the bizarre spinosaurs.

As these lineages were getting bigger, other groups like the coelurosaurs were getting smaller, so theropods occupied a wide variety of ecological roles. Many of the lineages that would come to rule the Cretaceous originated in the Jurassic, including dromaeosaurs, tyrannosaurs, and troodontids. Some of these coelurosaurs, which were fast runners with relatively long arms, started to use their feathered forelimbs in unique ways, experimenting with generating lift by flapping or gliding. It is now clear that active powered flight originated independently multiple times in these dinosaurs, but one particular dinosaur, *Archaeopteryx* (see pp.70–71), was especially adept at flying. Its relatives continue to rule the skies today.

▲ The time periods

Each geological period is introduced in a double-page feature, detailing the climatic conditions and, where relevant, plant life and prevalent animal forms. A timeline at the base of the pages indicates the chronological position of the period.

species name

syllable-by-syllable pronunciation guide

the name of the palaeontologist who first described the species, and the year in which the findings were published

habitat in which the animal lived

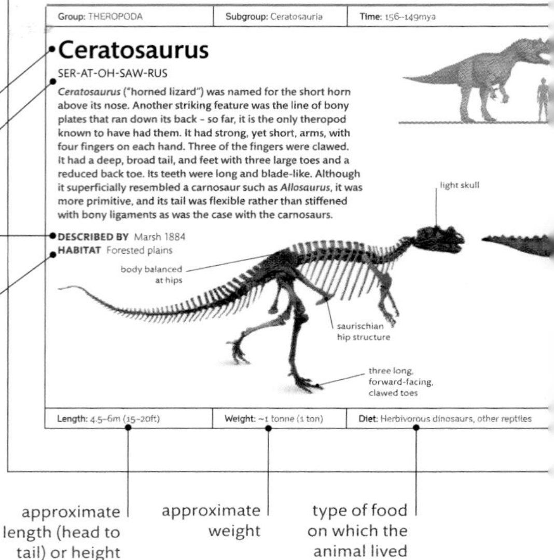

Group: THEROPODA | **Subgroup:** Ceratosauria | **Time:** 156–149mya

Ceratosaurus

SER-AT-OH-SAW-RUS

Ceratosaurus ("horned lizard") was named for the short horn above its nose. Another striking feature was the line of bony plates that ran down its back – so far, it is the only theropod known to have had them. It had strong, yet short, arms, with four fingers on each hand. Three of the fingers were clawed. It had a deep, broad tail, and feet with three large toes and a reduced back toe. Its teeth were long and blade-like. Although it superficially resembled a carnosaur such as *Allosaurus*, it was more primitive, and its tail was flexible rather than stiffened with bony ligaments as was the case with the carnosaurs.

DESCRIBED BY Marsh 1884

HABITAT Forested plains

Length: 4.5–6m (15–20ft) | **Weight:** ~1 tonne (1 ton) | **Diet:** Herbivorous dinosaurs, other reptiles

approximate length (head to tail) or height

approximate weight

type of food on which the animal lived

▼ Animal profiles

Arranged by time period and by group, each major animal is described in detail. The text includes information about the animal's physical features, diet, and lifestyle. Coloured bands summarize key information about classification, dates, dimensions, and diet. Each animal is illustrated by an artist's reconstruction and/or fossil remains. Key features are annotated. In each case a map identifies the approximate location of main fossil finds of that creature.

scientific name of main group

scientific name of subgroup

time when the animal existed in millions of years ago (mya) or years ago (ya)

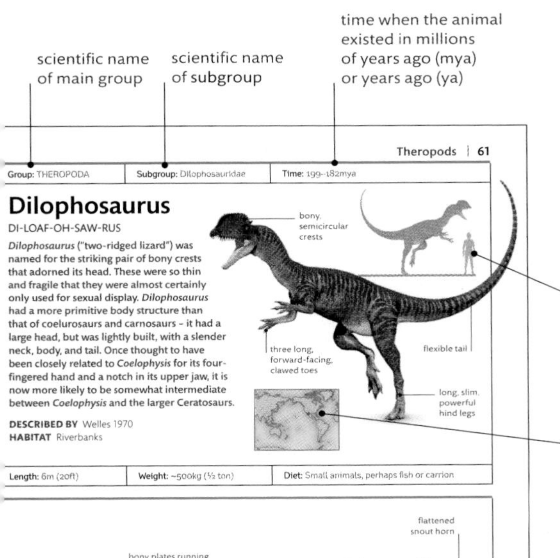

Theropods | 61
Group: THEROPODA
Subgroup: Dilophosauridae
Time: 199–182mya
Dilophosaurus
DI-LOAF-OH-SAW-RUS
Dilophosaurus ("two-ridged lizard") was named for the striking pair of bony crests that adorned its head. These were so thin and fragile that they were almost certainly only used for sexual display. *Dilophosaurus* had a more primitive body structure than that of coelurosaurs and carnosaurs – it had a large head, but was lightly built, with a slender neck, body, and tail. Once thought to have been closely related to *Coelophysis* for its four-fingered hand and a notch in its upper jaw, it is now more likely to be somewhat intermediate between *Coelophysis* and the larger Ceratosaurs.
DESCRIBED BY Welles 1970
HABITAT Riverbanks
bony, semicircular crests
three long, forward-facing, clawed toes
flexible tail
long, slim, powerful hind legs
Length: 6m (20ft)
Weight: ~500kg (½ ton)
Diet: Small animals, perhaps fish or carrion

COLOUR KEY

The bands at the top of each entry are colour-coded to identify the time period to which it belongs.

Time-Period colours

- Precambrian
- Cambrian
- Ordovician
- Silurian
- Devonian
- Carboniferous
- Permian
- Triassic
- Jurassic
- Cretaceous
- Paleogene
- Neogene
- Quaternary

the length or height of each animal is indicated in comparison to human dimensions

20cm (8in)

1.8m (6ft)

red dots indicate the location of principal fossil finds

▼ Feature spread

Animals of special interest are featured in a double-page entry. These show the animal in the type of landscape in which it is believed to have lived. Additional photographs and information boxes expand on the standard information for each featured animal.

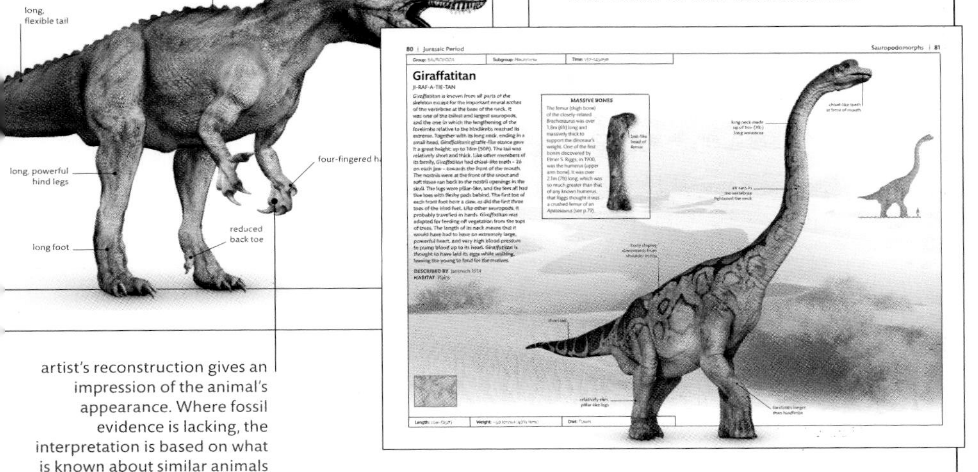

flattened snout horn
bony plates running along back
long, flexible tail
long, powerful hind legs
four-fingered h
reduced back toe
long foot
80 | Jurassic Period
Sauropodomorphs | 81
Giraffatitan
JI-RAF-A-TIE-TAN
MASSIVE BONES

artist's reconstruction gives an impression of the animal's appearance. Where fossil evidence is lacking, the interpretation is based on what is known about similar animals

CLASSIFICATION OF LIFE

AN ACCURATE UNDERSTANDING of living things and how they have evolved relies on their being classified into groups according to their similarity. Animals are classified in groups of decreasing diversity. The diagram outlines the evolution of the major groups of amniotes through time, from the diapsids and dinosaurs to the mammals. This is based on an analysis of shared features between species and their ancestors.

MONOPHYLETIC GROUPS

A monophyletic group is one that includes all the descendants of a common ancestor. This is in contrast to a paraphyletic group, where some descendants are left out, or a polyphyletic group, which artificially groups lineages that are not closely related. For example, dinosaurs are a monophyletic group only when birds – their direct descendants – are included. Palaeontologists sometimes use paraphyletic groups, such as "non-avian dinosaurs", for convenience, but with the understanding that these are not meaningful evolutionary groupings. In this book, the Jurassic and Cretaceous "Other Diapsid" sections are paraphyletic groupings, because they do not include dinosaurs.

KEY TO FOSSIL EVIDENCE

- Dinosaurs
- Pterosaurs
- Phytosaurs
- Pseudosuchians
- Rhynchosaurs
- Squamates
- Tanystropheids
- Ichthyopterygian
- Sauropterygians
- Diapsids
- Parareptiles
- Synapsids

NESTED CLADES

Palaeontologists represent the relationships between animals using cladograms, where lineages are represented by branches joined at nodes to show common ancestry. Each node corresponds to a clade and can be given a formal name. These clades can also be joined to each other to show broader groups. This means that a species can be part of many nested clades of broader and broader membership – similar to the old Linnaean System, but with more levels. Palaeontologists still use Linnaean names for genera and species, but most prefer to use nested clades instead of Linnaean ranks when discussing broader groups.

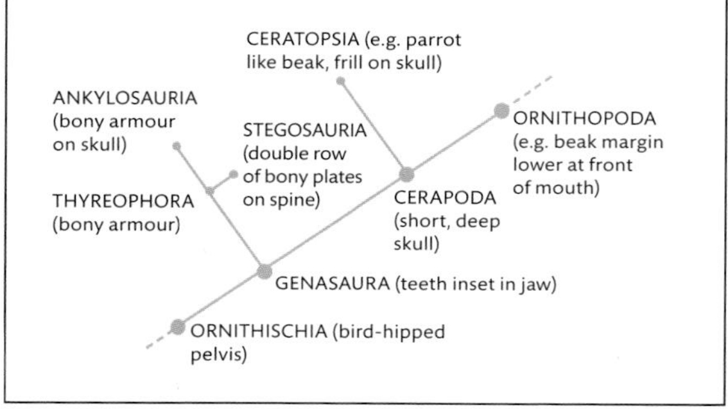

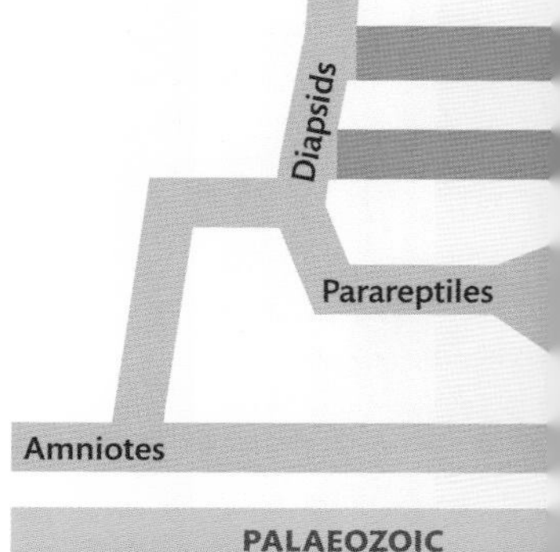

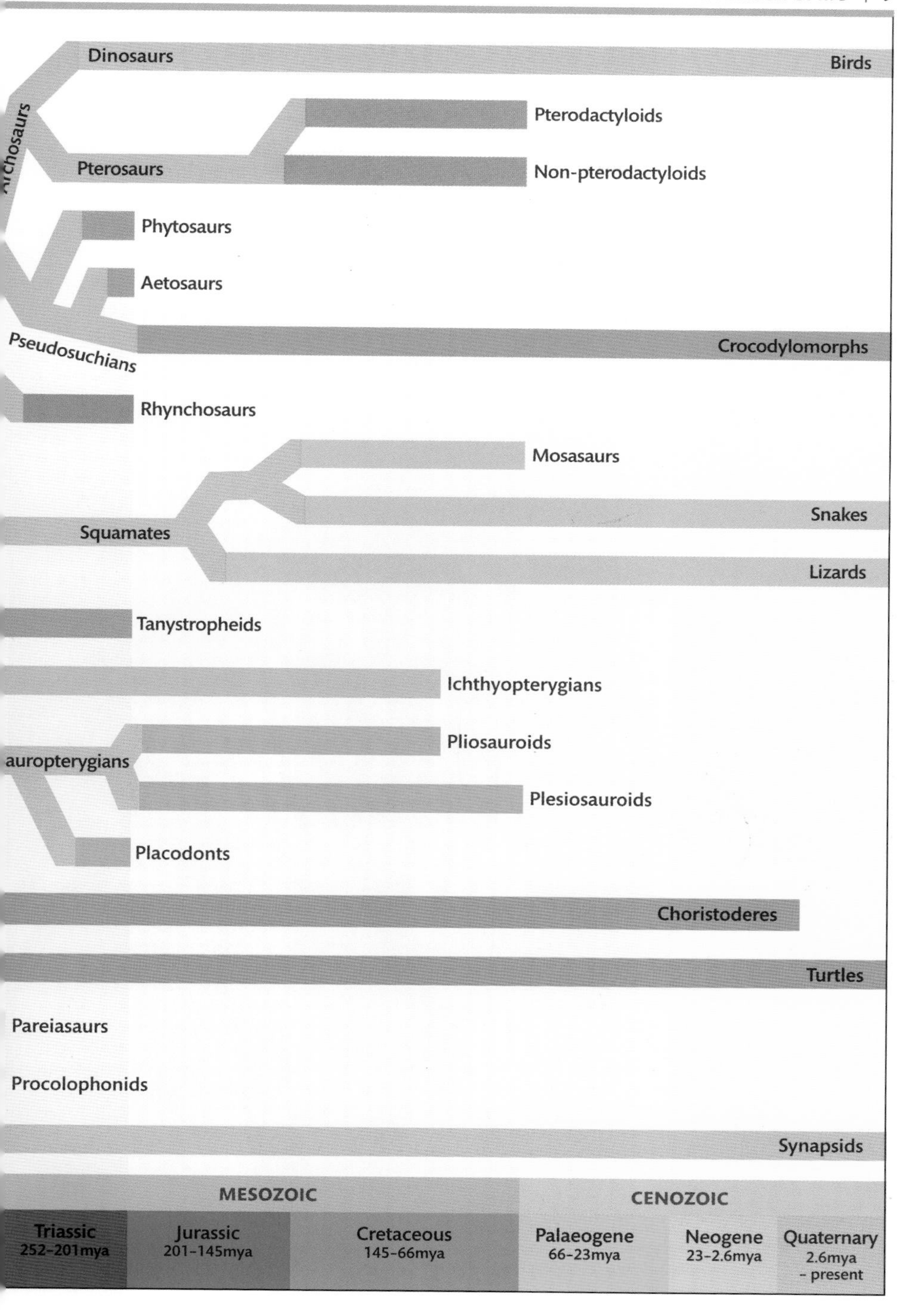
Dinosaurs
Birds
Archosaurs
Pterodactyloids
Pterosaurs
Non-pterodactyloids
Phytosaurs
Aetosaurs
Crocodylomorphs
Pseudosuchians
Rhynchosaurs
Mosasaurs
Snakes
Squamates
Lizards
Tanystropheids
Ichthyopterygians
Pliosauroids
auropterygians
Plesiosauroids
Placodonts
Choristoderes
Turtles
Pareiasaurs
Procolophonids
Synapsids
MESOZOIC
CENOZOIC
Triassic 252–201mya
Jurassic 201–145mya
Cretaceous 145–66mya
Palaeogene 66–23mya
Neogene 23–2.6mya
Quaternary 2.6mya – present

WHAT WERE DINOSAURS?

DINOSAURS DOMINATED the Earth's ecosystems for 165 million years. Although they evolved from a group of reptiles called the archosaurs, they possessed certain advanced features, as did other non-dinosaur groups such as the pterosaurs. Dinosaurs are defined as a group that shares certain characteristics. These include: upright limb posture (see opposite), three or fewer phalanges in the fifth finger of the hand, an elongate deltopectoral crest on the humerus, three or more sacral vertebrae, a ball-like head on the femur, and a fully-open acetabulum (hip-socket in the pelvis).

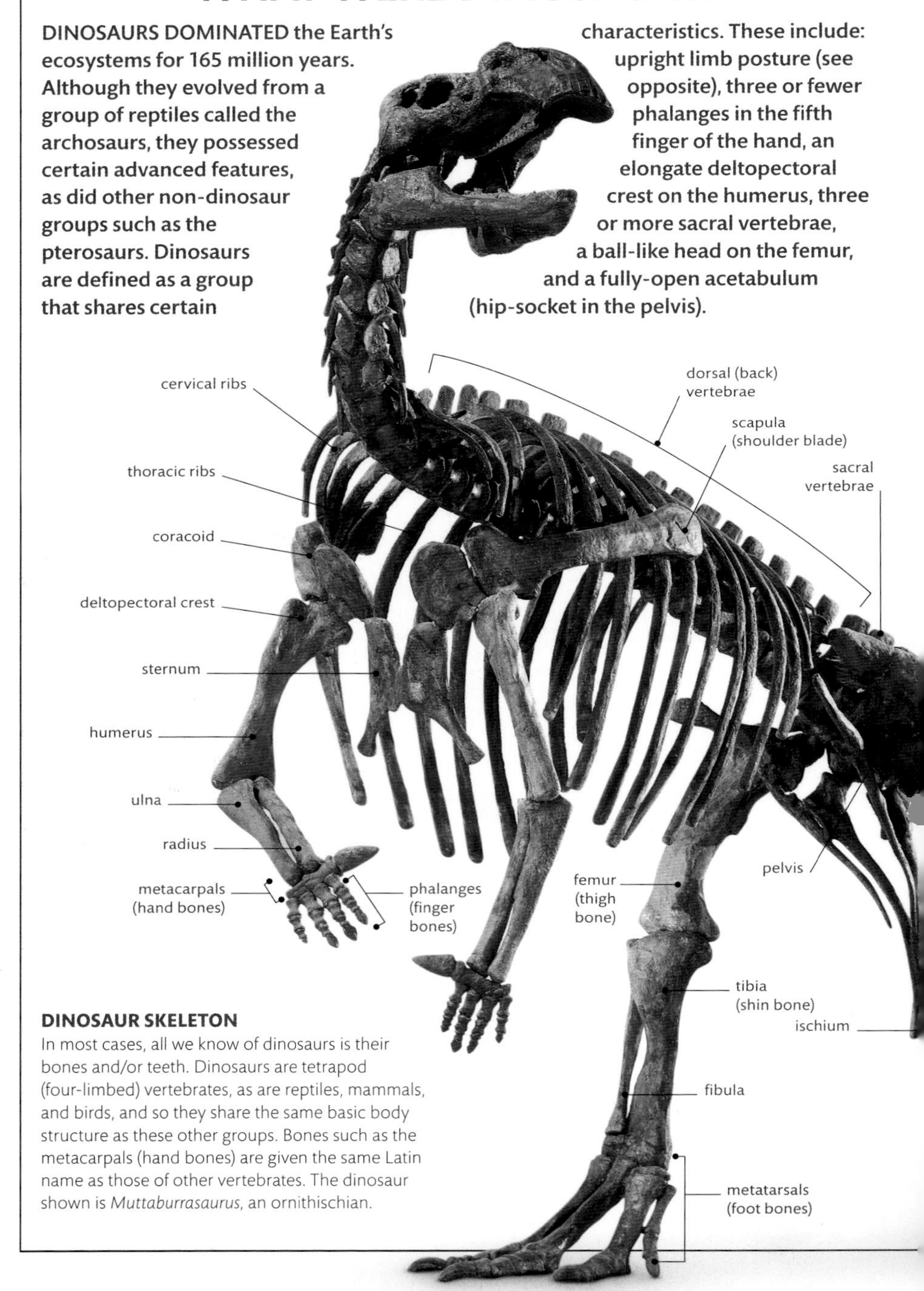

DINOSAUR SKELETON
In most cases, all we know of dinosaurs is their bones and/or teeth. Dinosaurs are tetrapod (four-limbed) vertebrates, as are reptiles, mammals, and birds, and so they share the same basic body structure as these other groups. Bones such as the metacarpals (hand bones) are given the same Latin name as those of other vertebrates. The dinosaur shown is *Muttaburrasaurus*, an ornithischian.

UPRIGHT POSTURE

Living lizards and crocodiles have a sprawling gait, with their knees and elbows held out at an angle from their bodies. One of the factors that led to the success of dinosaurs was their upright stance. This provides great advantages over the normal reptilian posture: it allows for longer strides, and therefore faster movement. Early meat-eating archosaurs and dinosaurs were often fast and agile hunters. The upright posture of the dinosaurs also allowed the evolution of bipedal (two-legged) walking.

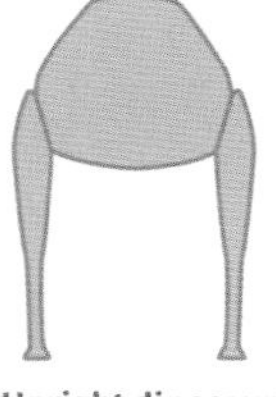

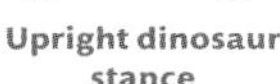

Upright dinosaur stance

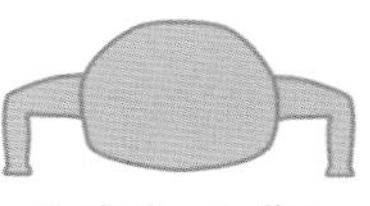

Typical sprawling lizard stance

ENDOTHERMIC?

The traditional view of sluggish, ectothermic (temperature changing) dinosaurs has largely given way to the theory that most were endothermic (constant temperature). Some were feathered, many were fast runners, and some lived in cold climates that would have been unsuitable for ectothermic animals.

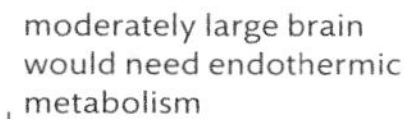

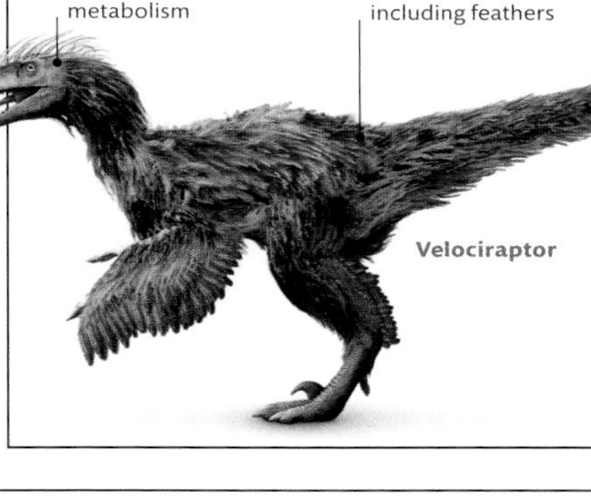

Velociraptor

DINOSAUR HIPS

In 1887, the English anatomist Harry Seeley recognized that there were two different types of dinosaur pelvis. Some dinosaurs had a typical lizard-like pelvic structure: Seeley called these saurischian ("lizard-hipped"). Another group had a pelvis that looked like that of modern birds. He called these ornithischian ("bird-hipped"). It is not clear whether the two groups evolved independently or from a common saurischian ancestor.

Saurischian pelvis

The pelvis is made up of the ilium, pubis, and ischium. In most saurischians, the ischium points backwards and the pubis points forwards.

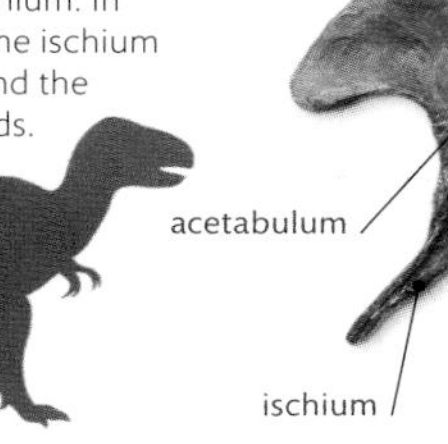

Ornithischian pelvis

In ornithischian dinosaurs, and some rare saurischian groups, the pubis lies against the ischium.

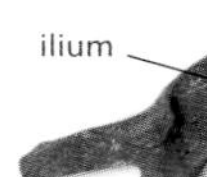

DINOSAUR AND BIRD EVOLUTION

THE DINOSAURS and their closest relatives were the dominant animals throughout the Mesozoic Era. The first step in the evolution of the dinosaurs occurred in the Permian Period. A new line of reptiles evolved, called the archosaurs ("ruling reptiles"). Some of these developed the ability to walk upright on two feet. The dinosaurs, crocodiles, and the pterosaurs all evolved from these early archosaurs during the Triassic Period. There are two main groups of dinosaurs: the saurischian ("lizard-hipped") group and the ornithischian ("bird-hipped") dinosaurs. Birds evolved from the theropod group of saurischian dinosaurs towards the end of the Jurassic Period. Together with the crocodiles, they are the only surviving archosaurs.

ORNITHISCHIANS

All of the ornithischians were plant-eaters, and they had special adaptations for herbivory. They had a special, toothless bone at the tip of the jaw called the predentary, which gave them a toothless beak in front of their teeth. They also had special adaptations to the jaw joint that allowed the teeth to grind against each other (occlude). The first major group to appear were the thyreophorans. This group of armoured dinosaurs included the stegosaurs and the ankylosaurs. Another group, the cerapodans, included the ceratopsians, the horned dinosaurs, the pachycephalosaurs, the bone-headed dinosaurs, and the ornithopods, a very diverse group of mainly bipedal dinosaurs.

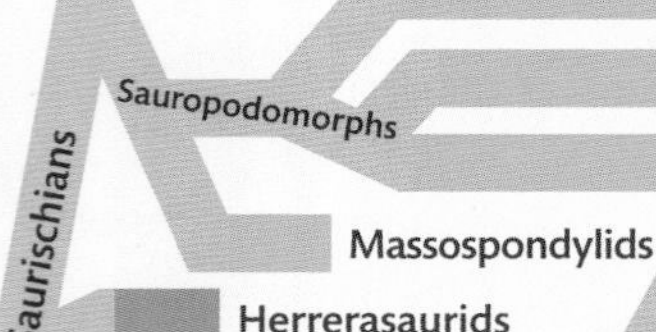

KEY TO FOSSIL EVIDENCE

- Aves
- Paravians
- Coelurosaurs
- Tetanurans
- Sauropodomorphs
- Saurischians
- Ornithischians

PALAEOZOIC		MESOZOIC	
Carboniferous 359–299mya	Permian 299–252mya	Triassic 252–201mya	Jurassic 201–145mya

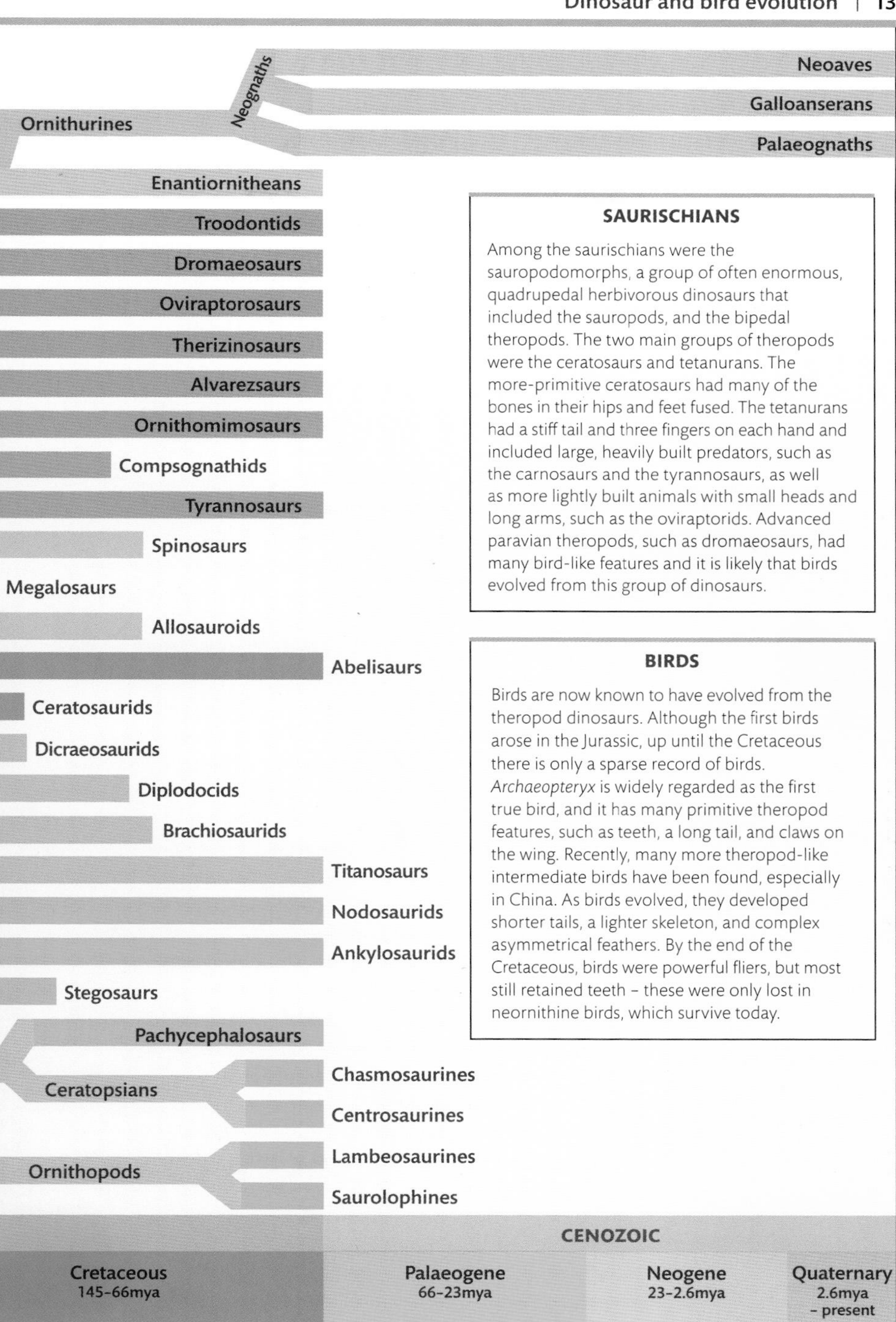

SAURISCHIANS

Among the saurischians were the sauropodomorphs, a group of often enormous, quadrupedal herbivorous dinosaurs that included the sauropods, and the bipedal theropods. The two main groups of theropods were the ceratosaurs and tetanurans. The more-primitive ceratosaurs had many of the bones in their hips and feet fused. The tetanurans had a stiff tail and three fingers on each hand and included large, heavily built predators, such as the carnosaurs and the tyrannosaurs, as well as more lightly built animals with small heads and long arms, such as the oviraptorids. Advanced paravian theropods, such as dromaeosaurs, had many bird-like features and it is likely that birds evolved from this group of dinosaurs.

BIRDS

Birds are now known to have evolved from the theropod dinosaurs. Although the first birds arose in the Jurassic, up until the Cretaceous there is only a sparse record of birds. *Archaeopteryx* is widely regarded as the first true bird, and it has many primitive theropod features, such as teeth, a long tail, and claws on the wing. Recently, many more theropod-like intermediate birds have been found, especially in China. As birds evolved, they developed shorter tails, a lighter skeleton, and complex asymmetrical feathers. By the end of the Cretaceous, birds were powerful fliers, but most still retained teeth – these were only lost in neornithine birds, which survive today.

MAMMAL EVOLUTION

MAMMALS ARE part of the synapsid lineage of amniotes, which arose during the Carboniferous period. There is debate about which particular groups of mammaliaformes can be considered true mammals, and so there is uncertainty about whether true mammals appeared in the Late Triassic or in the Jurassic. Throughout the Mesozoic, most mammals remained small, but we now know that some lineages diversified in their ecology and some even became large enough to prey on small dinosaurs. After the extinction of the dinosaurs, mammals took over global ecosystems, and today they are represented by three groups: placentals, marsupials, and monotremes.

SYNAPSIDS

The synapsids were a group of amniotes that dominated the land throughout the Permian and much of the Triassic. They had a single opening behind the eye socket on each side of the skull, giving them a more powerful bite. This group also evolved different types of teeth, a feature called heterodonty. These specialized teeth helped them process food in the mouth and improved their digestion. This, in turn, helped more specialized synapsids develop a faster metabolism, using more energy to grow more quickly and be more active. In the Triassic or Jurassic, mammals evolved from one of these specialized synapsid lineages called the cynodonts.

GENETIC EVIDENCE

For more than a century, palaeontologists used a classification system of mammals that was based on shared morphological characters, such as the shapes and features of the bones and teeth. This was revolutionized in the early 2000s by studies on the genetic relationships of living mammals using DNA and other molecules. These new techniques showed that many of the groups based on morphology were the result of convergent evolution, where two groups independently evolve the same features because they perform similar functions. Many of the discrepancies in these classification systems have since been solved, and palaeontologists now use a combined framework, considering both genetic and morphological data.

KEY TO FOSSIL EVIDENCE

- Placentals
- Marsupials
- Triconodontids
- Mammals
- Mammaliaformes
- Cynodonts
- Dicynodonts
- Sphenacodontid

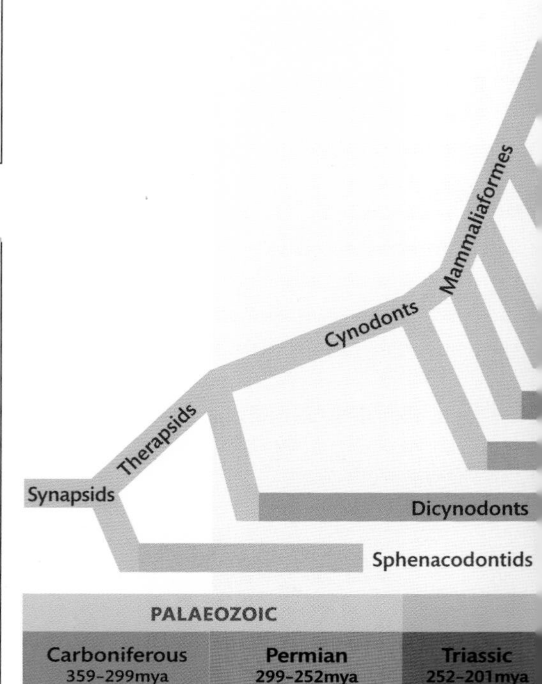

MAMMALIAFORMES

As palaeontologists discovered more cynodont fossils, the dividing line between mammals and their ancestors became blurred. Many of the close relatives of mammals had one or more of the defining mammal features, but not all of them. These might include features of the teeth or the specialized mammal-style middle ear. These new fossils now show that many of these features evolved several times independently, rather than in a straightforward way. Palaeontologists often call these almost-mammal species "proto-mammals" or lump them together in a paraphyletic group (see p.102) of "mammaliaformes"

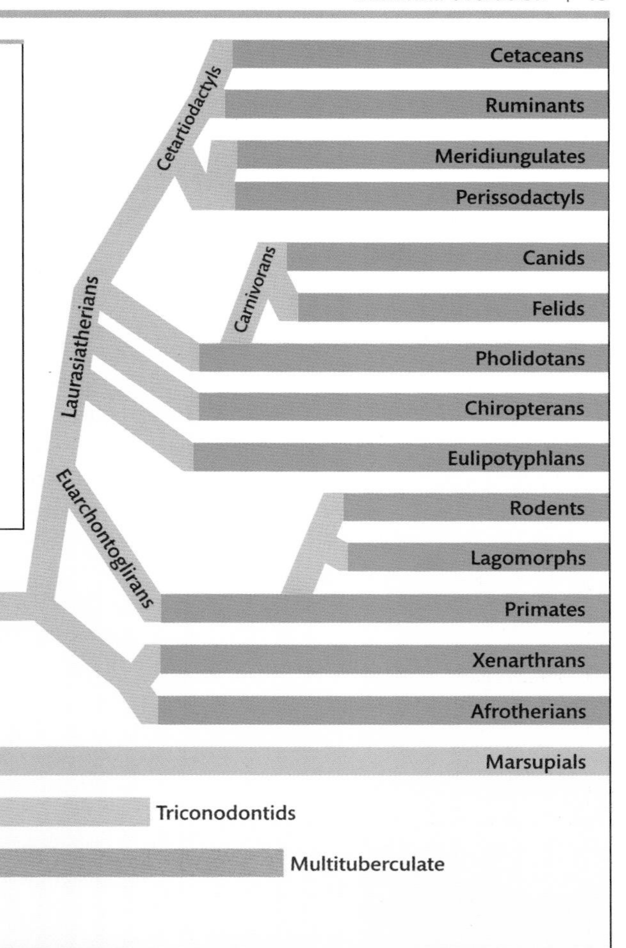

GEOLOGICAL TIME

GEOLOGISTS SUBDIVIDE the history of the Earth into very long time intervals called eons. Eons are, in turn, subdivided into eras, eras into periods, and periods into epochs.

The oldest rocks that have survived to be excavated were formed about four billion years ago, in the Archean Eon. The earliest fossils come from rocks of about this age.

ROCKS OF AGES

Where rock strata have remained undisturbed, a vertical section through the layers can reveal the rock types laid down during each time period. From these, it is possible to identify the environment (such as desert) and sometimes the age of fossils embedded in that rock, as shown below for rock strata at the Grand Canyon, USA.

ROCK	ENVIRONMENT	PERIOD
Shale, siltstone, mudstone	Tidal flat	Triassic
Limestone	Marine	Permian
Sandstone	Desert	
Shale	Savanna	
Mixed strata – shales, sandstones, limestones	Flood plain	Permian and Late Carboniferous
Limestone	Marine	Early Carboniferous
Limestone	Marine	Devonian
Limestone		Cambrian
Shale	Marine	
Sandstone	Marine	
Complex mixed strata	Marine and volcanic	Precambrian

Era	Dates	Period
CENOZOIC ERA	2.6mya–present	QUATERNARY
	23–2.6mya	NEOGENE
	66–23mya	PALEOGENE
MESOZOIC ERA	145–66mya	CRETACEOUS
	201–145mya	JURASSIC
	252–201mya	TRIASSIC
PALAEOZOIC ERA	299–252mya	PERMIAN
	359–299mya	CARBONIFEROUS
	419–359mya	DEVONIAN
	444–419mya	SILURIAN
	485–444mya	ORDOVICIAN
	542–485mya	CAMBRIAN
	4,600–542mya	PRECAMBRIAN

4,600mya 4,000mya 3,000mya

Grand Canyon
The Grand Canyon in Arizona, USA, provides one of the most impressive displays of sedimentary deposition of rocks. It is over 2,000m (6,000ft) deep, and shows a clear progression over 300 million years. Layers include sandy limestone, petrified sand dunes, and shale.

Macrauchenia

The ice ages of the Quaternary Period led to the evolution of many mammals adapted to cold climates, such as mammoths. Modern humans evolved.

In the Neogene, large expanses of grassland spread, inhabited by grazing mammals and predatory giant birds. The first hominids evolved from primate ancestors.

Titanis

Ambulocetus

After the end-Cretaceous mass extinctions, mammals evolved into large forms. Some took to a marine way of life. Giant flightless birds evolved.

The Cretaceous was a time of flowering plants, browsing duck-billed dinosaurs, immense tyrannosaurid predators, armoured ankylosaurs, and horned ceratopsians.

Didelphodon

Barosaurus

In Jurassic times, the land was dominated by huge sauropods and large predators. A wide variety of pterosaurs evolved. Mammals remained small.

The Triassic marked the dawning of the age of dinosaurs. Advanced synapsids died out after giving rise to mammals.

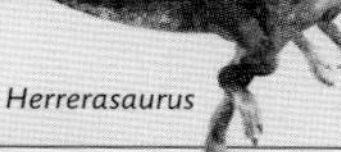
Herrerasaurus

Dimetrodon

Synapsids became the dominant land animals during the Permian. The period ended with the largest mass extinction event ever.

Tropical forests flourished and oxygen levels were very high during this period. The first reptiles moved onto the land.

Graeophonus

Eastmanosteus

The Devonian Period was a time of rapid evolution. Ammonoids and bony fish evolved and diversified. Trees appeared on land, as did insects and the first four-limbed animals.

Invertebrate species recovered rapidly during the Silurian Period. Primitive lycopods and myriapods became the first true land animals.

Sagenocrinites

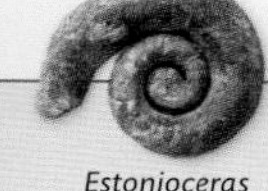
Estonioceras

Ordovician seas teemed with primitive fish, trilobites, corals, and shellfish. Plants made their first approaches onto land. The period ended with mass extinctions.

The first animals with skeletons and hard parts such as shells and exoskeletons evolved during the Cambrian "explosion of life". The oceans teemed with trilobites, brachiopods, and the first jawless fish.

Xystridura

Charniodiscus

During the Precambrian, life arose in the oceans – first as single-celled bacteria and algae, then evolving into soft-bodied, multicellular animals, such as jellyfish and worms.

2,000mya | 1,000mya | 500mya | 250mya | 0

FOSSIL EVIDENCE

EVIDENCE OF prehistoric life comes from remains (such as bones) that over time have become mineralized to form fossils. Most animal fossils are of creatures that died in or near water. After death, if the hard remains, such as teeth and bones, quickly become covered in mud or sand, dissolved minerals in water are able to seep into the bone pores, initiating a process called permineralization. This reinforces the bones and makes them harder. Sometimes minerals replace the bone completely, petrifying it, or minerals dissolve the bone away, leaving a bone-shaped hollow called a mould.

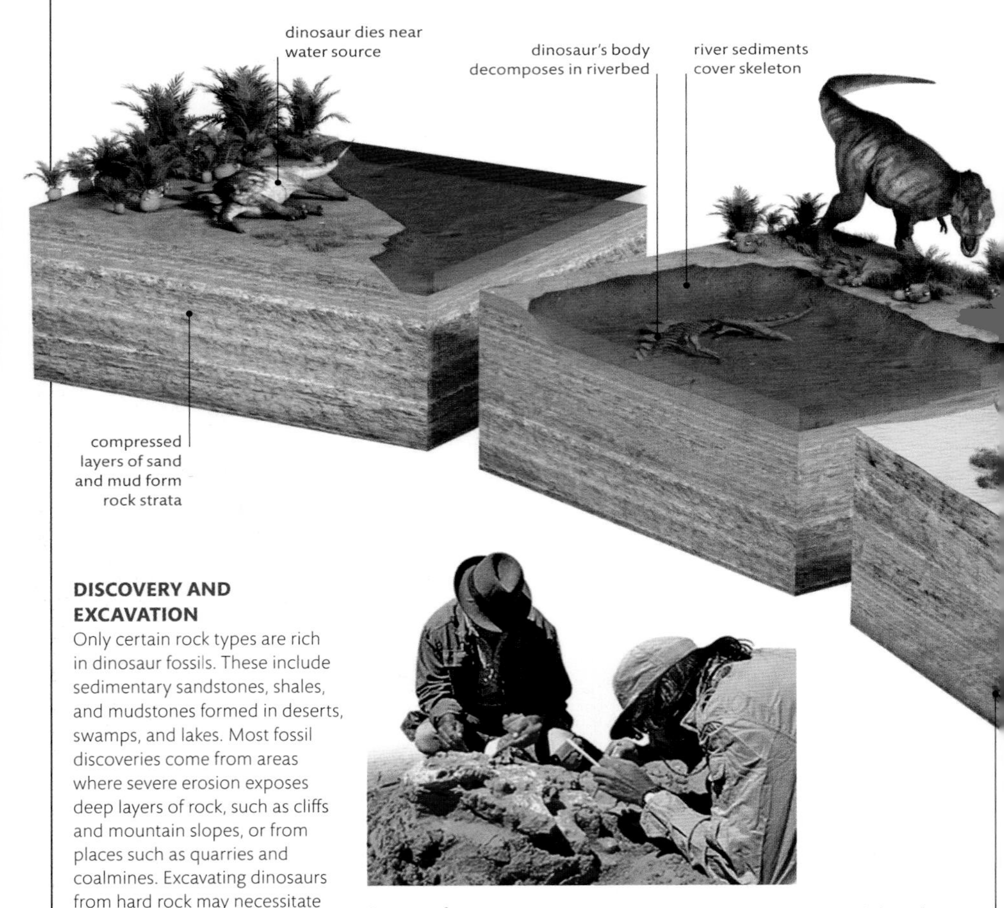
dinosaur dies near water source
dinosaur's body decomposes in riverbed
river sediments cover skeleton
compressed layers of sand and mud form rock strata
skeleton becomes mineralized, enabling it to withstand weight of rock forming overhead

DISCOVERY AND EXCAVATION

Only certain rock types are rich in dinosaur fossils. These include sedimentary sandstones, shales, and mudstones formed in deserts, swamps, and lakes. Most fossil discoveries come from areas where severe erosion exposes deep layers of rock, such as cliffs and mountain slopes, or from places such as quarries and coalmines. Excavating dinosaurs from hard rock may necessitate the use of power tools or explosives. Fossils in desert areas can sometimes be exposed by carefully brushing away the sand covering them.

Excavation
Palaeontologists dig around bones to estimate the fossil's state of preservation and size. Delicate scraping and chiselling is needed to reveal delicate structures without causing damage.

TRACE FOSSILS

Besides fossil remains such as bones, skin impressions, and teeth, dinosaurs left other clues to their existence and lifestyle. Trace fossils include fossil footprint trackways made in mud that dried out in the sun (below right). Coprolites (fossilized droppings) give further clues about a dinosaur's anatomy and lifestyle. Piles of gizzard stones (stones that were swallowed to aid digestion) are also rarely found.

Fossil of the future?
Dinosaur fossils found in the Sahara Desert were formed when the area was a swamp. This camel skeleton may become fossilized if it is buried by the sand.

The fossilization process

Fossilization depends on the rapid covering of remains by sediments that exclude oxygen and thereby arrest the normal processes of decay. In time, minerals from the covering sediments permeate the bone or other tissues, replacing the original material. The soft tissues usually decompose before the process of fossilization begins, so soft-bodied animals, such as jellyfish and worms, are generally poorly preserved.

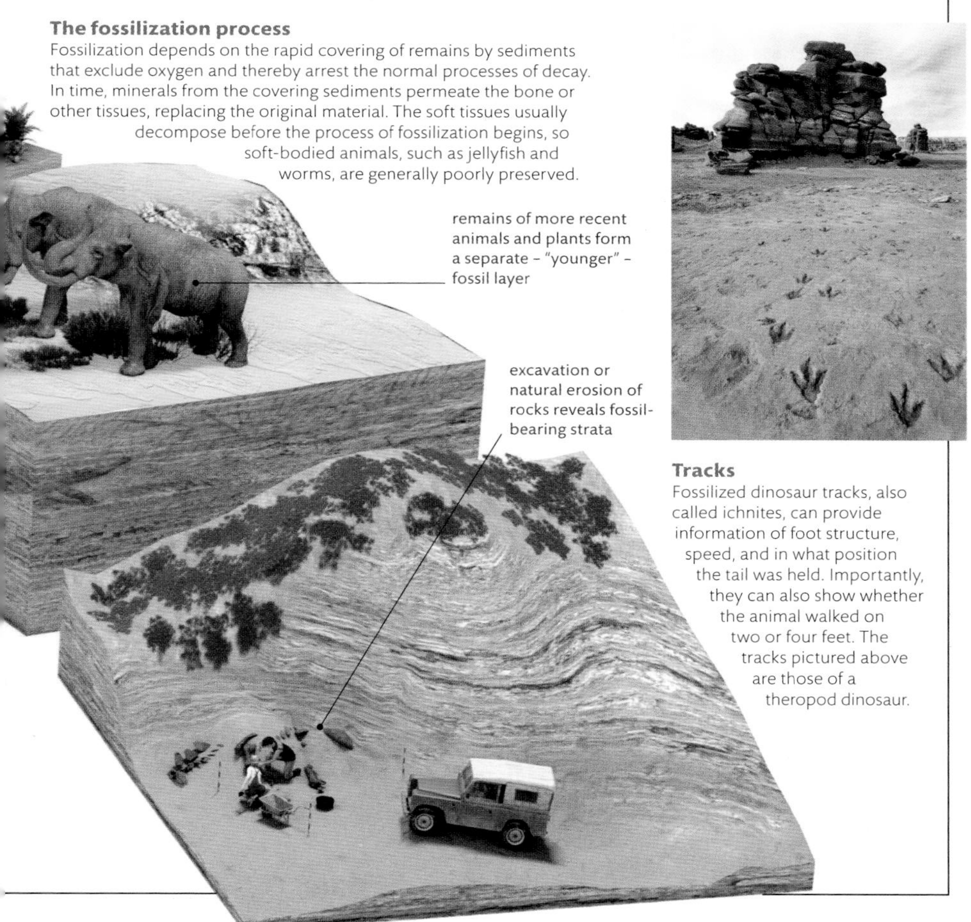
remains of more recent animals and plants form a separate - "younger" - fossil layer
excavation or natural erosion of rocks reveals fossil-bearing strata

Tracks
Fossilized dinosaur tracks, also called ichnites, can provide information of foot structure, speed, and in what position the tail was held. Importantly, they can also show whether the animal walked on two or four feet. The tracks pictured above are those of a theropod dinosaur.

PRECAMBRIAN TIME

4,600–542 MILLION YEARS AGO

CURRENT ESTIMATES put the age of the Earth at 4,600 million years. The years from that date up until about 542 million years ago (mya) are grouped in one large division of geological time – the Precambrian. This was the time when life first evolved, although evidence of living creatures is sparse. During the first million years of its existence, the world was hot, molten, and totally inhospitable to life. As the earth cooled, volcanic gases and water vapour formed a crude atmosphere, and the oceans began to form. Before about 3,000mya, most of the Earth's surface was volcanic rock, with craters and many unstable regions of volcanic activity. Stable continental areas then began to form. Fossil evidence of life is first seen in rocks dated at 4,200 million years old. The earliest life forms on Earth were simple cells and bacteria. More complex cells and algae first evolved about 1,600mya. Multicelled plants appeared about 850mya, whereas the earliest animals (sponges) are about 635–660 million years old.

Extraordinary fossils found in Precambrian rocks of the Ediacaran Period date from 600mya, and are thought to be the remains of the earliest multicellular animals (*Spriggina* and *Charniodiscus*, opposite). The *Mawsonites* fossil (above) has been interpreted as a jellyfish or traces of a primitive worm.

4,600mya | 4,000mya | 3,000mya

PRECAMBRIAN LANDMASSES

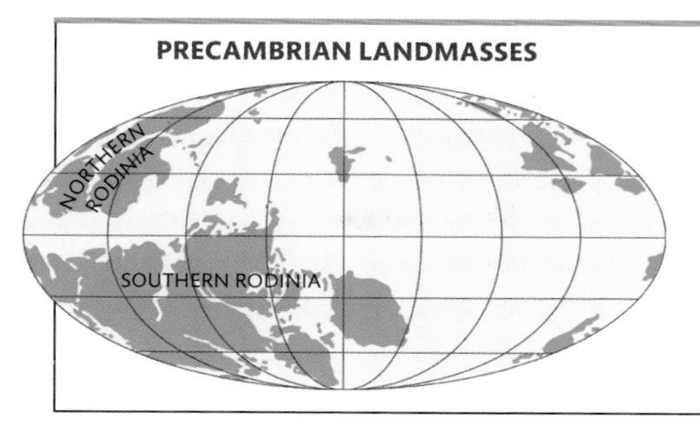

Landmasses began to form about 3,000mya. About 900mya they joined to form the first supercontinent, Rodinia. By 750mya, Rodinia was sitting across the Equator, but it then started to move southwards. This triggered the Cryogenian period. Between 750 and 650mya, Rodinia split into two halves (see map), but by the end of the Precambrian it again formed a single continent called Pannotia.

PRECAMBRIAN LIFE

The first living cells were microscopic organisms possibly living in hot springs. By about 3,500mya, photosynthesizing algae may have formed layered structures called stromatolites. As oxygen built up in the atmosphere, multicellular animals evolved.

Stromatolites
These layered silica or limestone structures were created by colony-forming algae and provide evidence of early life. *Collenia* (left) was a type of Precambrian stromatolite.

First animals
Late in the Precambrian, the first true animals and plants appeared. *Spriggina* (above) was a strange, long, tapering animal, with V-shaped segments running along the length of its body.

Filter feeder
The feather-shaped fossil of *Charniodiscus* is thought to represent an early filter-feeding animal that lived on the sea floor late in the Precambrian.

CAMBRIAN PERIOD

542–485 MILLION YEARS AGO

THE BEGINNING of the Cambrian Period marks the start of the Phanerozoic Era, known as the "Era of Abundant Life". The Cambrian is remarkable for the amazing increase in the number of marine animals and plants. However, the land remained barren. The evolutionary spurt was probably helped by the relatively warm climate. Rapid continental movements led to high sea levels and large expanses of shallow-water environments covered large parts of the continents. These shallow seas were warm worldwide as there were no ice caps at either pole.

Burgess Shale
A huge variety of fossil animals, called the Burgess Shale fauna, have been found in shale rocks of the Cambrian Period in British Columbia, Canada. About 180 species of invertebrates have been discovered there. Chordates (creatures with a primitive backbone called a notochord), such as *Pikaia* (right), are also preserved in the Burgess Shale.

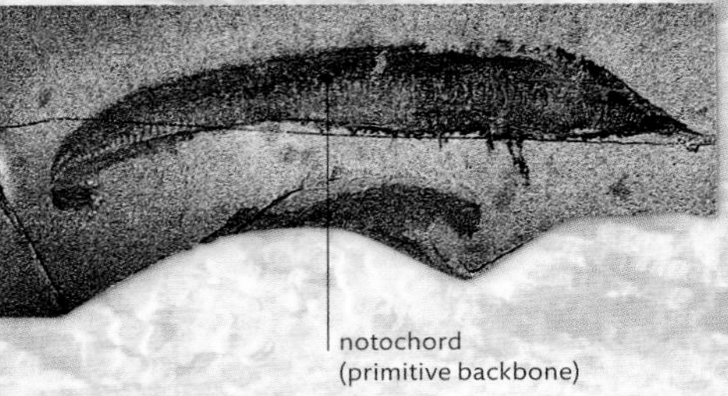

4,600mya | 4,000mya | 3,000mya

CAMBRIAN LANDMASSES

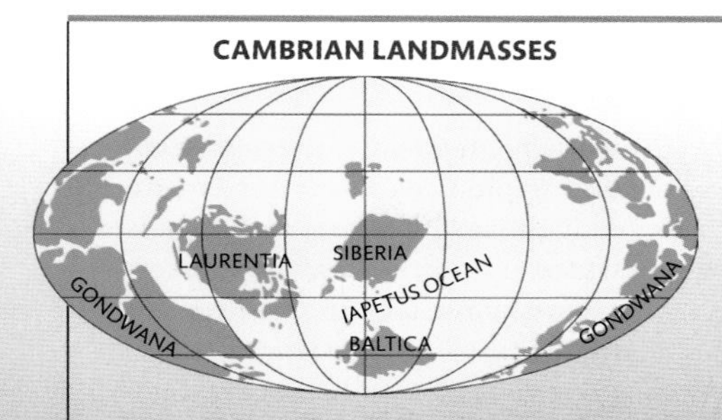

Continental drift was remarkably rapid during the Cambrian. The supercontinent Pannotia broke up, and the continents moved apart (see map). This caused a rise in sea levels. By 500mya, the continents of Laurentia, Baltica, and Siberia were lined up along the Equator, with the supercontinent Gondwana (present-day South America, Africa, Antarctica, Australia, and Asia) extending into temperate regions.

CAMBRIAN LIFE

Almost all of the present-day major groupings of animals started to appear during this almost 60 million-year period, including worms, crabs, shellfish, and sponges. Many of the new animals had a hard external skeleton. This provided protection within which multicellular organisms could grow.

Arthropods

This group of early animals were remarkable for being the first organisms with eyes – compound eyes rather like those of insects. *Xystridura* (left), like other arthropods, lived on the sea bed, and had a skeleton composed of a head shield, a jointed thorax, and a tail shield.

jointed bony thorax with many legs

many small protective plates

Wiwaxia

Many mollusc- and worm-like animals evolved. *Wiwaxia* (right), a mollusc-like creature covered with hard spines, lived on the ocean floor.

ORDOVICIAN PERIOD

485–444 MILLION YEARS AGO

THE ORDOVICIAN PERIOD saw the first movement of life from the oceans onto the land. Before about 450mya, the land was barren apart from mats of algae close to shores. After this time, early forms of liverwort-type plants evolved, possibly from larger algae. They quickly colonized boggy regions and areas around lakes and ponds. In the oceans, the earliest jawless fish (agnathans) appeared, which were mainly tear-drop-shaped animals covered in bony plates. Towards the end of the Ordovician, two extinction events occurred about a million years apart. The first of these was caused by global warming, which resulted in the retreat of glaciers. Warm-water coral reefs died out, and three-quarters of all marine species became extinct within less than a million years.

Brachiopods
Mainly living attached to reefs, these were the most abundant two-shelled organisms in the oceans. *Strophomena* (left) was a medium-sized brachiopod.

4,600mya	4,000mya	3,000mya

ORDOVICIAN LANDMASSES

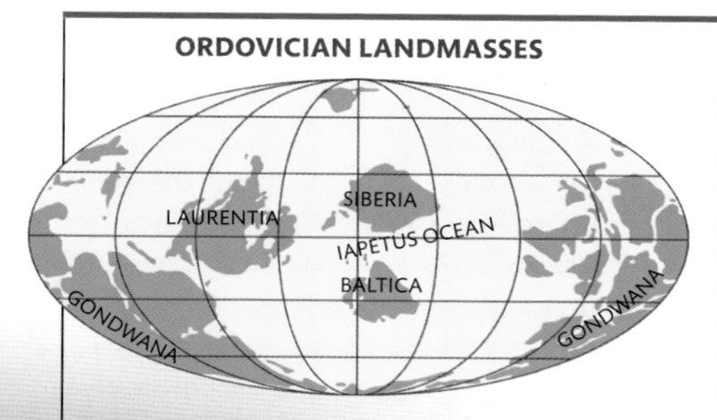

At the beginning of the Ordovician Period, the supercontinent Gondwana still lay in the southern hemisphere. The other continents were spread out along the Equator and were gradually pushed apart by the expansion of the Iapetus Ocean (see map). Later in the Ordovician Period, the movement of Gondwana towards the South Pole triggered another extinction event. By 440mya, present-day North Africa lay over the South Pole.

ORDOVICIAN LIFE

The warm equatorial seas that existed for much of the Ordovician were ideal for the evolution of marine life. Coral reefs appeared, and soon spread widely. The oceans were also filled with jellyfish, sea anemones, and other colony-forming organisms. A variety of shelled animals, such as brachiopods, lived on sea beds around the reefs.

fan-like colonies held in branching, tube-like structures

Graptolites

Graptolites, such as *Rhabdinopora* (above), were unusual colony-forming organisms that floated through the water feeding through minute tentacles.

Nautiloids

Primitive shelled cephalopods (a group that includes modern cuttlefish) called nautiloids were abundant. Many, such as *Estonioceras* (left), were coiled. They moved by squirting water out of a tube in their body cavity.

segmented body

Trilobites

The new coral reefs were an ideal habitat for trilobites. *Sphaerexochus* (above) was a typical Ordovician trilobite, with a body of 11 segments.

2,000mya 1,000mya 500mya 250mya 0

SILURIAN PERIOD

444–419 MILLION YEARS AGO

BEFORE THE SILURIAN PERIOD, the land had been largely barren apart from some mossy growth and liverworts close to water. The landscape was transformed by the appearance of the first true land plants. These were primitive and simple, but had branching stems, roots, and tubes for water transport. Such plants were able to grow to larger sizes than their more primitive antecedants. As areas around water began to be covered by vegetation, soil built up, and water became trapped in the land. These conditions allowed the first advances of animal life onto the land. By the end of the Silurian, there were many land arthropods, including primitive centipedes, and spider- and scorpion-like arachnids.

SILURIAN LIFE

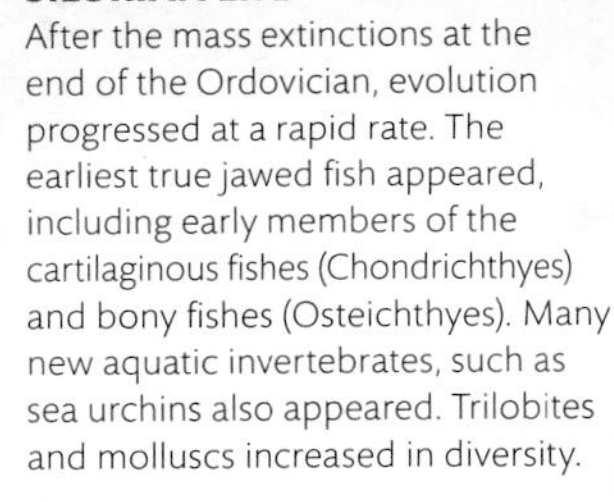

After the mass extinctions at the end of the Ordovician, evolution progressed at a rapid rate. The earliest true jawed fish appeared, including early members of the cartilaginous fishes (Chondrichthyes) and bony fishes (Osteichthyes). Many new aquatic invertebrates, such as sea urchins also appeared. Trilobites and molluscs increased in diversity.

Sea scorpion
The largest Silurian marine invertebrates were the sea scorpions (eurypterids), such as *Pterygotus* (above). In some cases longer than a man, these were the dominant hunters of the early seas, and a few may have been able to crawl ashore. Eurypterids were cousins of horseshoe crabs and arachnids, but are not directly related to scorpions.

4,600mya	4,000mya	3,000mya

SILURIAN LANDMASSES

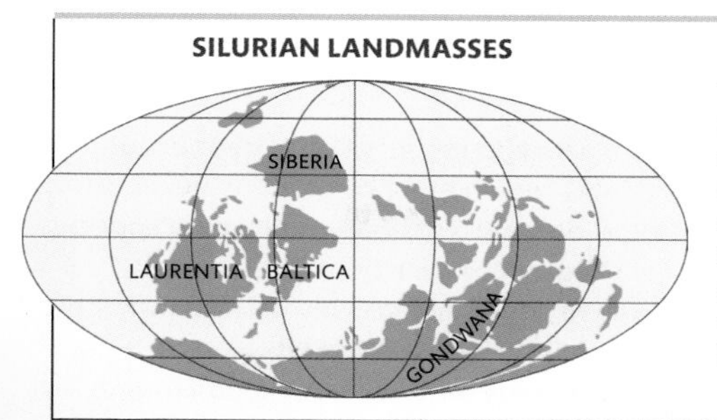

The second Ordovician extinction event led to rising sea levels and flooding of some low-lying areas. The climate became warmer and less seasonal. The supercontinent Gondwana was still lying over the South Pole, with Laurentia sitting astride the Equator (see map). The collision of smaller landmasses formed new mountain ranges. By the end of the Silurian, all of the landmasses were grouped closely together.

characteristic Y-shaped branching stems

Early plants

The earliest plants, such as *Cooksonia* (left), had a simple branching form. They had water-conducting vessels, and reinforced stems.

freely moving arms

Crinoids

Despite their plant-like appearance, crinoids (sea lilies) were animals related to starfish and sea urchins. Many species populated Silurian seabeds. *Sagenocrinites* (above) was a small one, with a very compact head and many slender tentacles.

Birkenia

Despite the evolution of the jawed fish, jawless fish (agnathans) were still thriving. *Birkenia* (right) was a freshwater fish that probably fed on the algae that it strained out of muddy lake- and riverbeds.

2,000mya 1,000mya 500mya 250mya 0

DEVONIAN PERIOD

419–359 MILLION YEARS AGO

DURING THE DEVONIAN green plants of increasing size started to spread across the landscape. Later in the period, some plants evolved woody tissue and the first trees appeared. Towards the end of the Devonian, the climate became much warmer. Droughts were common and alternated with periods of heavy rain. Sea-levels fell worldwide, and large deserts formed. A wide variety of fish evolved – the Devonian is often called "the Age of Fish". Swampy deltas and river estuaries provided an important habitat for the emergence of animal life onto land.

Prototaxites
Although they looked like tree trunks, *Prototaxites* (right) were actually a type of fungus. Growing up to 8m (26ft) tall, they were the largest organisms on land during the Devonian.

Life on land
About 370mya, the first four-legged vertebrates, such as *Acanthostega* (right), ventured out of the water onto the land. They evolved from one of the groups of lobe-finned fish, whose living relatives are lungfish. The paired fleshy fins of these fish evolved into limbs.

sharp teeth suggest a diet of fish

DEVONIAN LANDMASSES

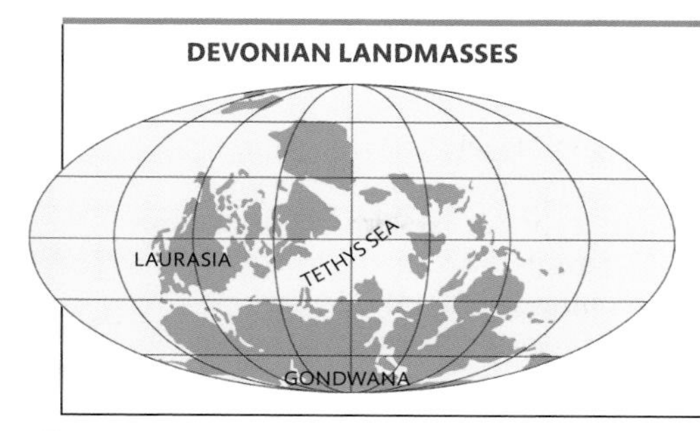

At the start of the Devonian the northern landmasses formed one large supercontinent called Laurasia, separated from Gondwana by the Tethys Sea (see map). Huge mountain ranges formed in the future eastern North America and western Europe. By the mid Devonian, Gondwana and most of Laurasia had moved south of the Equator. The lands that would become China and Siberia were north of the Equator.

DEVONIAN LIFE

The most important evolutionary step during the Devonian was the development of four-limbed land animals. Arthropods also moved onto the land and the first insects appeared. The oceans teemed with armoured jawless fish and the more modern jawed fish. Sharks and ammonoids (a group of molluscs) were common.

Lungfish

Lungfish, such as *Dipterus* (right) have primitive lungs as well as gills, and survive during seasons of drought by breathing air in watertight burrows in the mud.

body armour

Placoderms

Abundant in Devonian seas, placoderms were jawed fish. Towards the end of the Devonian some of these creatures reached 8m (26ft) in length. *Eastmanosteus* (above), although less than 2m (6½ft) long, was a fearsome hunter.

Early sharks

The seas of the Late Devonian teemed with squid, small fish, and crustaceans, which were ideal prey for early sharks such as *Cladoselache* (left). This shark did not have scales on its body.

CARBONIFEROUS PERIOD

359–299 MILLION YEARS AGO

THE CARBONIFEROUS PERIOD opened with tropical conditions over much of the Earth. There were large coastal seas and vast swamps covering the coastal plains. Insects and amphibians found the swamps an ideal environment. Early amphibians looked rather like salamanders, but they soon evolved into many forms, including giant ones. In the warm, moist climate, huge forests of giant tree ferns flourished, producing an oxygen-rich environment. There were also giant horsetails, clubmosses, and seed-bearing plants. Large amounts of decaying vegetation led to the build up of thick layers of peat, which would later become coal deposits.

CARBONIFEROUS LIFE

There were many types of amphibian tetrapods, ranging from newt-like animals to those the size of crocodiles. By the late Carboniferous, the first reptiles appeared, with the evolution of a shelled egg. The early reptiles were small, but they spread rapidly on land, moving into drier, upland areas.

diamond-shaped leaf bosses

Trees

Huge forests of woody trees, such as *Lepidodendron* (a specimen of fossilized bark is shown left), spread worldwide. This giant clubmoss had a column-like trunk with skeletal, spreading branches and could grow to heights of 35m (115ft), with a trunk diameter of over 2m (6½ft).

First reptile?

When first discovered, *Westlothiana lizziae* (right) was hailed as the first real reptile. It is now regarded as reptile-like rather than a true reptile. *Westlothiana* has features that indicate an evolutionary position between primitive tetrapods and true reptiles.

Insects and arthropods

Already populous by the Carboniferous, insects and arthropods continued to diversify. *Graephonus* (above) was a whip scorpion, with six legs and a pair of pincers.

4,600mya | 4,000mya | 3,000mya

CARBONIFEROUS LANDMASSES

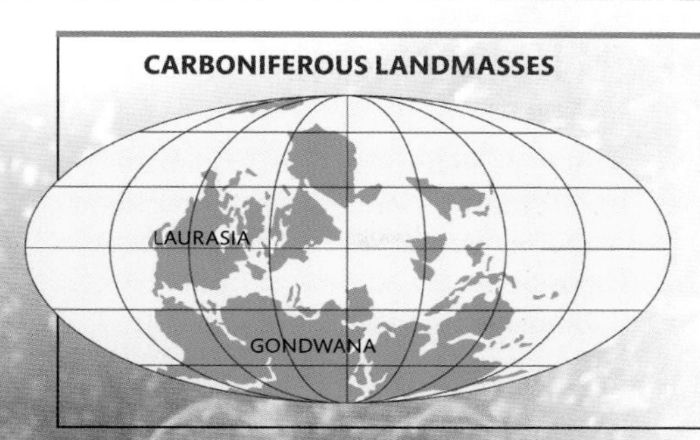

At the beginning of the Carboniferous, most of the Earth's landmasses were arranged in two great supercontinents: Laurasia in the north and Gondwana in the south (see map). Later in the Carboniferous they moved closer together. Gondwana started to move over the South Pole again. The advance and retreat of the ensuing icesheets led to at least two ice ages in the last half of the Carboniferous.

Life in a swamp

The spread of the forests raised oxygen levels around the world, and this factor, combined with the increasingly humid conditions, may have allowed the evolution of very large amphibians and insects. Crocodile-like amphibians, such as *Eryops*, hunted at the bottom of the swamps. The largest Carboniferous dragonflies had wingspans of up to 75cm (2½ft), and some early scorpions were more than 60cm (2ft) long.

early reptiles, such as **Hylonomus** migrated to drier areas

the dragonfly **Meganeura**, one of the largest flying insects

Eryops, a large, amphibian aquatic hunter

PERMIAN PERIOD

299–252 MILLION YEARS AGO

THE PERMIAN was a time of dramatic climatic changes. At the start of the Permian, Gondwana was still in the grip of an ice age. It gradually warmed over the next few million years, as it moved northwards. Large parts of Laurasia became very hot and dry, and massive expanses of desert formed. This had a damaging effect on amphibian populations: they were confined to the fewer damp areas and many species became extinct. This provided the opportunity for reptiles to spread more widely and diversify. Continental upheavals and further extreme climate changes at the end of the Permian led to the largest mass extinction event ever. More than half of all animal families became extinct.

PERMIAN LIFE
Reptiles continued to spread rapidly. Among the dominant land animals of the time were the synapsids (mammal-like reptiles). By the mid Permian, these had diversified into the therapsids. Later in the period, a mammal-like group of therapsids called the cynodonts ("dog teeth") evolved.

Pareiasaurs
These were large, primitive, herbivorous reptiles. *Elginia* (a fossil skull is shown left) was one of the smaller pareiasaurs and one of the last of the group.

Early parareptile
Procolophon (right) was a member of the Parareptilia – a clade of reptiles. It may have eaten insects. Later members of its family were larger, and had teeth designed for eating plants.

4,600mya 4,000mya 3,000mya

PERMIAN LANDMASSES

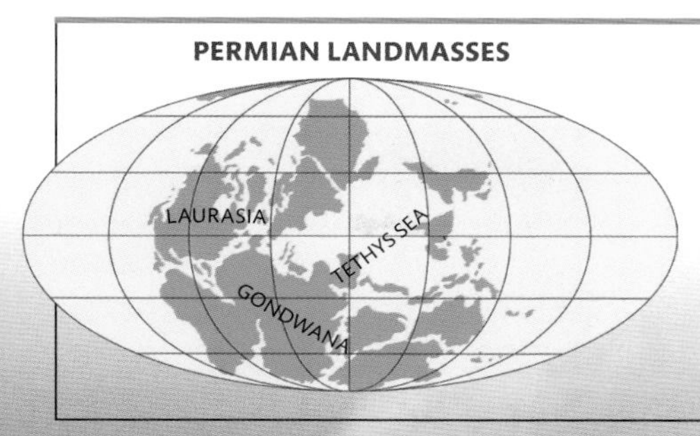

Throughout the Permian Period, Laurasia in the north and Gondwana in the south continued to move closer together (see map). By the end of the period, they collided to form the giant supercontinent of Pangaea. This straddled the Equator, and was bounded to the east by the shallow Tethys sea. Many shallow coastal seas disappeared, inland desert areas became much larger, and sea levels fell worldwide.

Synapsids
One of the best-known of this group of early Permian predators, *Dimetrodon* (left) had a distinctive sail on its back, which was supported by bony rods. This is thought to have been for heat exchange or sexual display.

large, sharp teeth indicate carnivorous lifestyle

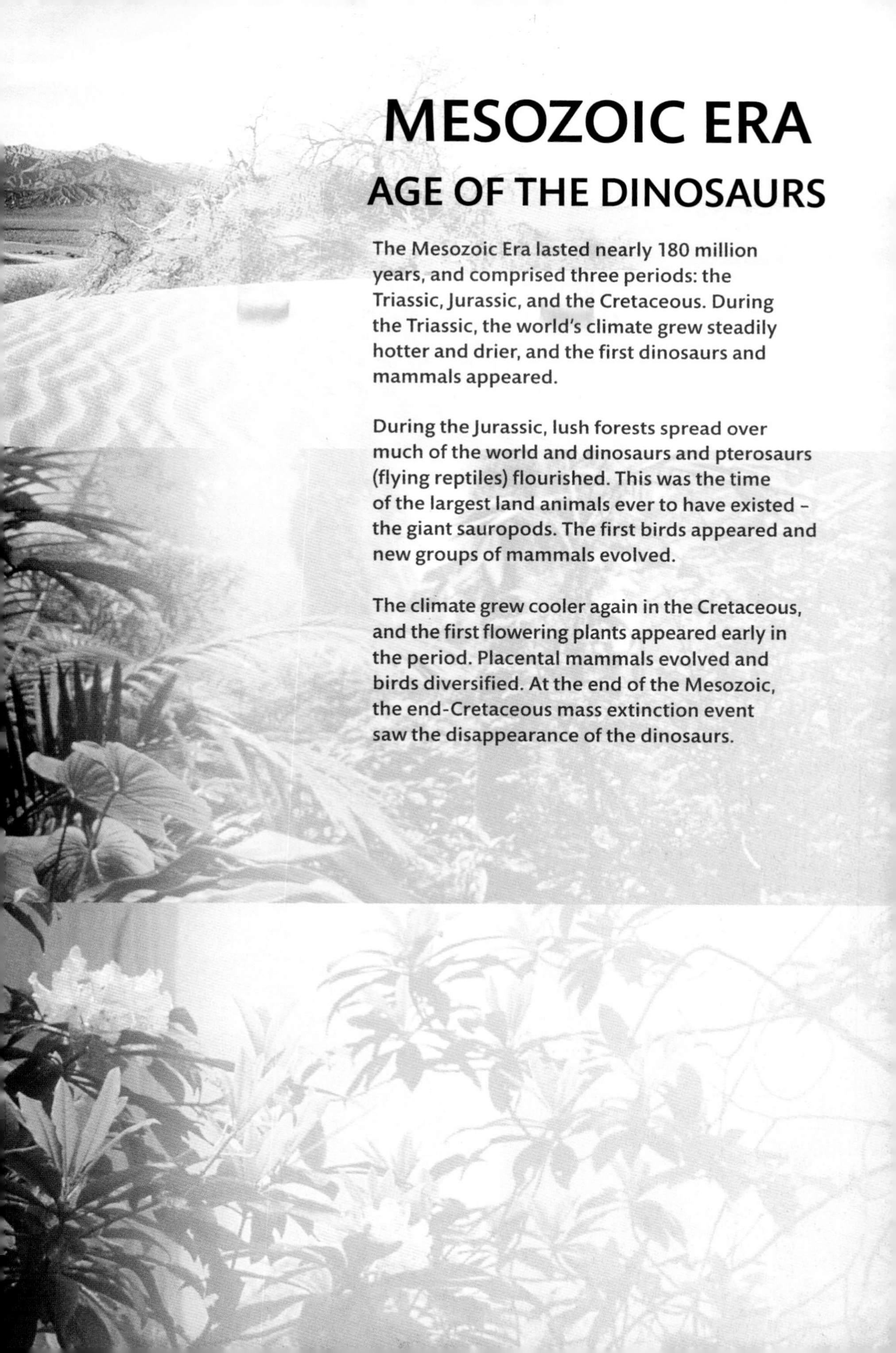

MESOZOIC ERA

AGE OF THE DINOSAURS

The Mesozoic Era lasted nearly 180 million years, and comprised three periods: the Triassic, Jurassic, and the Cretaceous. During the Triassic, the world's climate grew steadily hotter and drier, and the first dinosaurs and mammals appeared.

During the Jurassic, lush forests spread over much of the world and dinosaurs and pterosaurs (flying reptiles) flourished. This was the time of the largest land animals ever to have existed – the giant sauropods. The first birds appeared and new groups of mammals evolved.

The climate grew cooler again in the Cretaceous, and the first flowering plants appeared early in the period. Placental mammals evolved and birds diversified. At the end of the Mesozoic, the end-Cretaceous mass extinction event saw the disappearance of the dinosaurs.

TRIASSIC PERIOD

252–201 MILLION YEARS AGO

AFTER THE MASS extinctions at the end of the Permian, the oceans and land were left startlingly empty of life. It took about 10 million years for the Earth's ecosystems to recover. During the Triassic, the surviving reptile groups spread widely and the first true dinosaurs appeared. A large area of the world's landmasses lay in the equatorial belt, and interior regions were subject to alternating seasons of heavy monsoonal rains and drought. The climate was generally warm, and there were no polar ice caps. These conditions favoured certain types of plants, such as seed ferns and conifers, which could cope with arid conditions, and horsetails in damper areas. The climate grew even drier at the end of the Triassic.

TRIASSIC LIFE

Most of the synapsids, which had dominated the land in Permian times, did not survive into the Triassic. The surviving groups radiated again, but lost a number of their ecological niches to new reptile groups – the archosaurs ("ruling reptiles") and the rhynchosaurs (a short-lived group of diapsid reptiles). Several lines of aquatic reptiles also evolved. These were the nothosaurs, placodonts, and ichthyosaurs. The Late Triassic saw the arrival of the dinosaurs, crocodilians, pterosaurs, turtles, and primitive mammals.

saurischian ("lizard-hipped") pelvic structure

Cynodonts

The cynodonts (dog-like mammal ancestors) were a group that survived into the Triassic. *Cynognathus* (above) was a large Triassic carnivorous cynodont.

First dinosaurs

The earliest dinosaurs evolved about 230mya. *Herrerasaurus* (right) and *Eoraptor*, both found in Argentina, are thought to be among the earliest dinosaurs. Both of them were agile bipedal hunters.

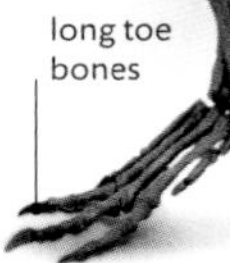

long toe bones

4,600mya 4,000mya 3,000mya

TRIASSIC LANDMASSES

The supercontinent Pangaea reached its maximum state of fusion in the mid Triassic about 230mya. Parts of present-day Asia may have formed islands, but most of the landmasses of the Earth were in contact. Pangaea straddled the Equator, reaching from pole to pole. The temperature gradient between the Equator and the poles was far less extreme than it is in modern times, and there were no polar ice caps.

Flying reptiles

The earliest known flying reptiles are pterosaurs, found in Late Triassic deposits. They were accomplished flyers with wings made of skin attached to an elongated fourth finger. *Peteinosaurus* (above) was an early pterosaur (a primitive form) with a short neck, and a long bony tail.

Triassic vegetation

Many plants of the Palaeozoic Era relied on high rainfall and damp conditions. *Dicroidium* (right) was a seed fern the size of a small tree. It thrived in swampy areas of the Triassic southern hemisphere. In many areas, however, such tropical vegetation was gradually replaced by plants, such as cycads, ginkgos, and evergreen trees, which were better suited to the prevailing arid conditions.

DIAPSIDS

AFTER THE end-Permian extinction, surviving diapsids slowly recovered and began to occupy some of the ecological roles previously filled by synapsids.

Most major diapsid groups had originated prior to the extinction, but in the Triassic they underwent significant radiations. Early in the Triassic, some diapsids returned to the sea, leading to the ichythyosaur and sauropterygian lineages, but the origins of these groups are still a mystery. Placodonts were among the successful colonizers of the oceans, but they would not survive the Triassic–Jurassic extinction. They developed turtle-like shells to protect them from the carnivorous sauropterygians, but they were only distantly related to turtles.

Lepidosaurs diversified, too, and early in the Triassic the first squamates (the group to which lizards and snakes belong) appeared. The most successful Triassic diapsids were the archosauromorphs, including pseudosuchians and dinosaurs, alongside some groups with no modern descendants, such as pterosaurs, phytosaurs, and rhynchosaurs.

Pterosaurs, the first vertebrates to evolve powered flight, took to the skies early in the Triassic. Pseudosuchians, the broad group that gave rise to modern crocodiles, were the dominant terrestrial animals for most of the Triassic, and they developed a staggering diversity that included top carnivores, lumbering herbivores, and agile bipeds. For reasons still unclear, however, they were slowly displaced by dinosaurs towards the end of the Triassic. The first dinosaurs were small carnivores, but by the end of the Triassic there were also large herbivorous sauropodomorphs, such as *Plateosaurus*.

Group: SAUROPTERYGIA	Subgroup: Placodontia	Time: 237–227mya

Henodus

HEN-OH-DUS

Henodus ("single-tooth") was an armoured placodont shaped rather like a modern turtle. It was as wide as it was long, and its back and belly were protected by bony, many-sided plates that made up a defensive shell. This in turn was completely covered with plates of tough horn. *Henodus* lived in near-shore marine environments. It had a strangely square snout and one tooth per side of the mouth. It also had a horny beak at the front of the mouth, similar to that of modern turtles. *Henodus* scraped plants off the bottom using suction and filter feeding. Its short, clawed feet may have been webbed.

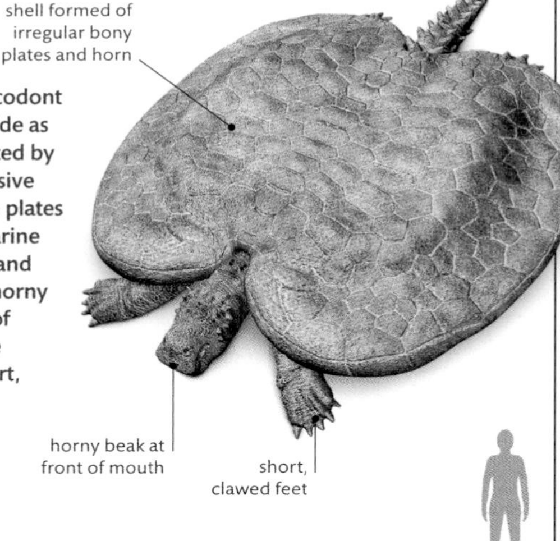

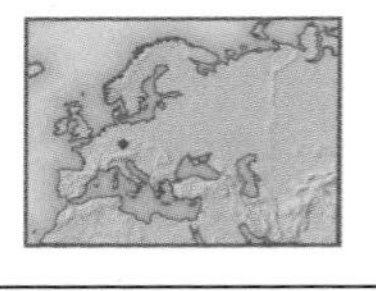

DESCRIBED BY
Huene 1936
HABITAT
Lagoons

Length: 1m (3¼ft)	Weight: Not calculated	Diet: Herbivore or filter feeder

Group: SAUROPTERYGIA	Subgroup: Placodontia	Time: 243–235mya

Placodus

PLAK-OH-DUS

This placodont shows few adaptations for an aquatic lifestyle, with a stocky body, short neck, and sprawling limbs. However, there were webs of skin between the toes and the tail was flattened from side to side. There may also have been a fin on the tail. *Placodus* ("flat tooth") had forward-pointing teeth used to pluck shellfish off rocks. Flat teeth on the palate met other teeth on the lower jaw to produce an efficient crushing action.

DESCRIBED BY Agassiz 1843
HABITAT Seashores

flat teeth

Lower jaw

Upper jaw

sprawling, five-toed feet

belly ribs formed protective shell

Length: 2m (6½ft)	Weight: Not calculated	Diet: Shellfish, crustaceans

Group: SAUROPTERYGIA	Subgroup: Placodontia	Time: 227–201mya

Psephoderma

SEF-OH-DER-MAH

This relatively well-known placodont was remarkably turtle-like in appearance. Its body was broad and flat and covered with hexagonal plates. Its limbs were paddle-shaped. *Psephoderma* ("pebble skin") had great biting power – a horny beak at the front of its mouth could pluck shellfish, which were then crushed by its teeth and jaws.

DESCRIBED BY von Meyer 1858
HABITAT Shallow seas

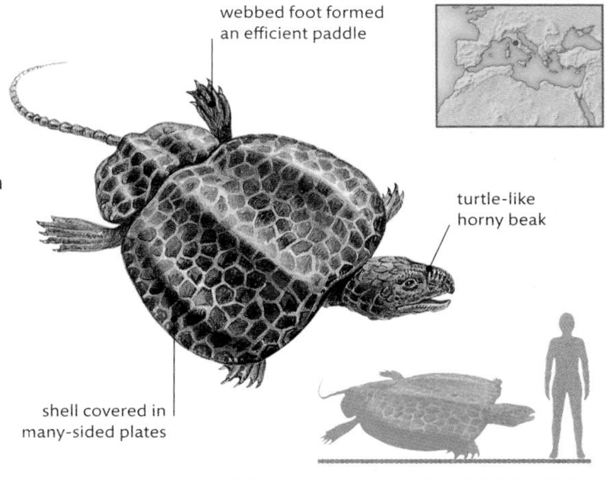

Length: 1.8m (6ft)	Weight: Not calculated	Diet: Shellfish, crustaceans

Group: SAUROPTERYGIA	Subgroup: Nothosauria	Time: 242mya

Ceresiosaurus

SER-EE-SEE-OH-SAW-RUS

Ceresiosaurus ("lizard of Lugano") had toes that were much longer than those of other nothosaurs due to hyperphalangy (an increased number of bones in each toe). The length of its feet, which may have been webbed, meant that efficient swimming flippers were formed. The forelegs were longer than the hind legs, suggesting that they were the ones used most for steering. *Ceresiosaurus* swam using a combination of its large front flippers for propulsion and undulation of the body.

DESCRIBED BY Peyer 1931
HABITAT Shallow seas

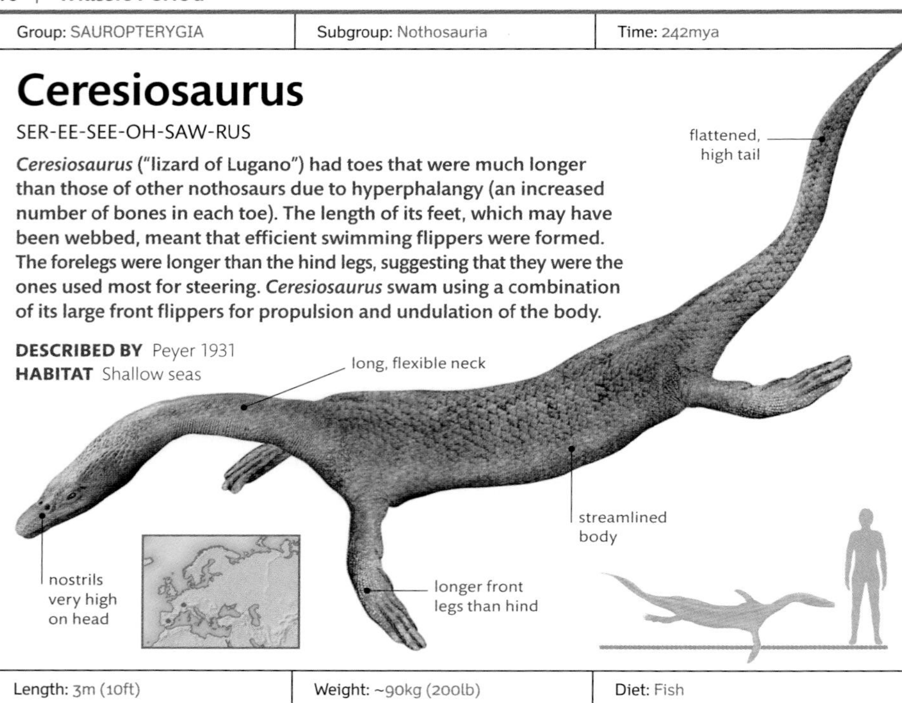

Length: 3m (10ft)	Weight: ~90kg (200lb)	Diet: Fish

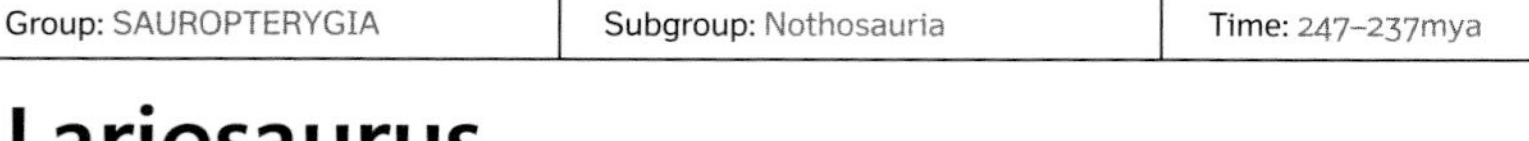

Group: SAUROPTERYGIA	Subgroup: Nothosauria	Time: 247–237mya

Lariosaurus

LA-REE-OH-SAW-RUS

Lariosaurus was a small member of the nothosaur group of marine reptiles. It had primitive adaptations to an aquatic lifestyle. Its neck and toes were very short, and the webs of skin on its hindfeet would have been small, and therefore not much use for fast swimming. Its front feet formed paddle-like flippers. In common with other members of its group, it had flexible knee and ankle joints. Fossil embryos preserved with an adult *Lariosaurus* show that it gave birth to live young, an adaptation that would have allowed it to reproduce without returning to land.

DESCRIBED BY
Curioni 1847
HABITAT
Coastal shallows

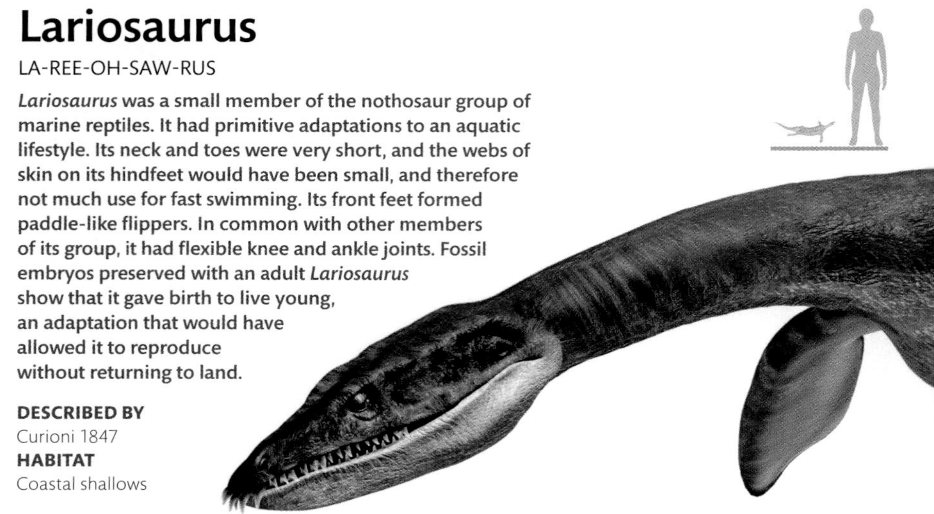

Length: 60cm (24in)	Weight: ~10kg (22lb)	Diet: Small fish, shrimp

Group: SAUROPTERYGIA	Subgroup: Nothosauria	Time: 240–210mya

Nothosaurus

NOH-THO-SAW-RUS

Nothosaurus **("false lizard") was a typical nothosaur. Its long body, neck, and tail were flexible and moderately streamlined. Some fossils have impressions of webbed skin between the five toes of the feet. The head was slim, and the jaws contained many sharp, thin teeth that interlocked when the mouth was closed. The nostrils were placed high on the head, close to the eyes.**

DESCRIBED BY
Münster 1834
HABITAT
Coastal regions

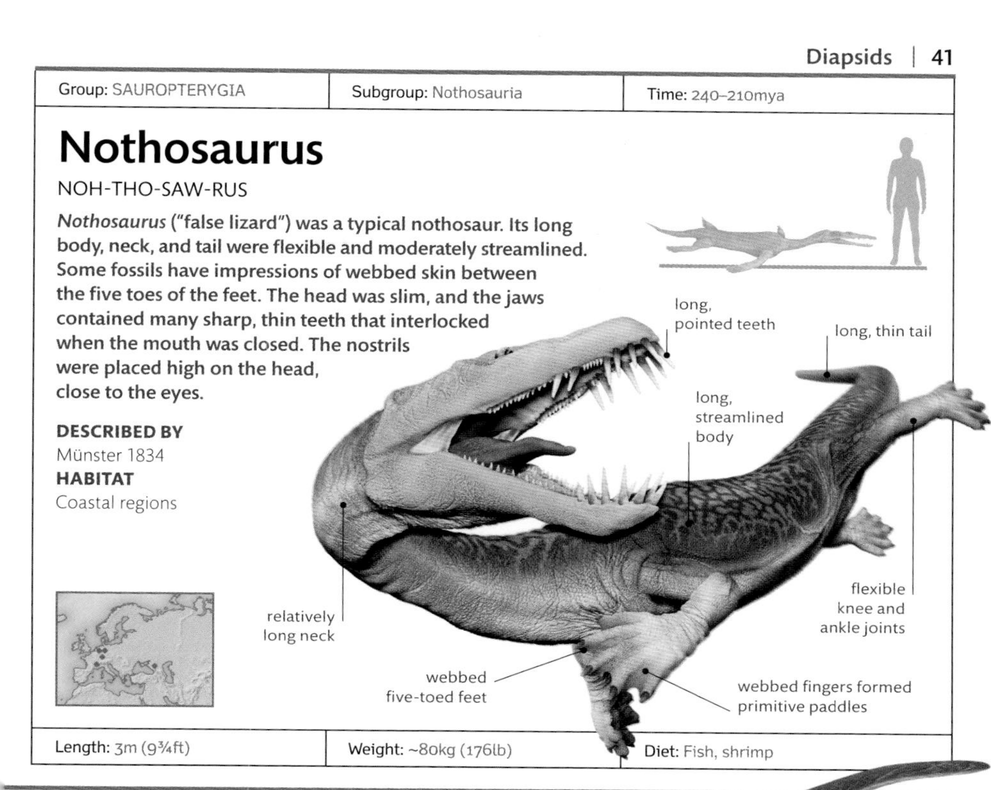

Length: 3m (9¾ft)	Weight: ~80kg (176lb)	Diet: Fish, shrimp

Group: ICHTHYOPTERYGIA	Subgroup: Ichthyosauria	Time: 247–242mya

Mixosaurus

MIX-OH-SAW-RUS

This "mixed reptile" is thought to have been an intermediate form between primitive ichthyosaurs and more advanced types. It had a fish-like body, typical of advanced ichthyosaurs, with a dorsal fin on its back. However, it only had a small fin on the top of its tail. Its paddles were short, with the front pair longer than the hind ones.

DESCRIBED BY Baur 1887
HABITAT Oceans

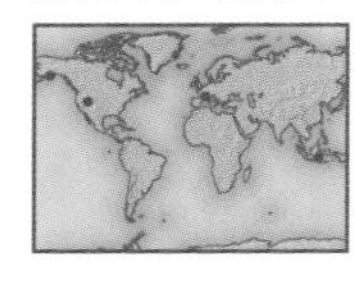

Length: 1m (3¼ft)	Weight: Not calculated	Diet: Fish

Group: ICHTHYOPTERYGIA	Subgroup: Ichthyosauria	Time: 237–210mya

Shonisaurus

SHON-EE-SAW-RUS

***Shonisaurus* ("Shoshone mountain reptile") is the largest ichthyosaur known. It had the typical ichthyosaur shape, with its head and neck, body, and tail making up equal thirds of its length. However, it had several features that indicate that it was an offshoot from the main ichthyosaur line. Its jaws were very long, and had teeth only at the front. Its paddles were also unusually long, and of equal length.**

DESCRIBED BY Camp 1976
HABITAT Oceans

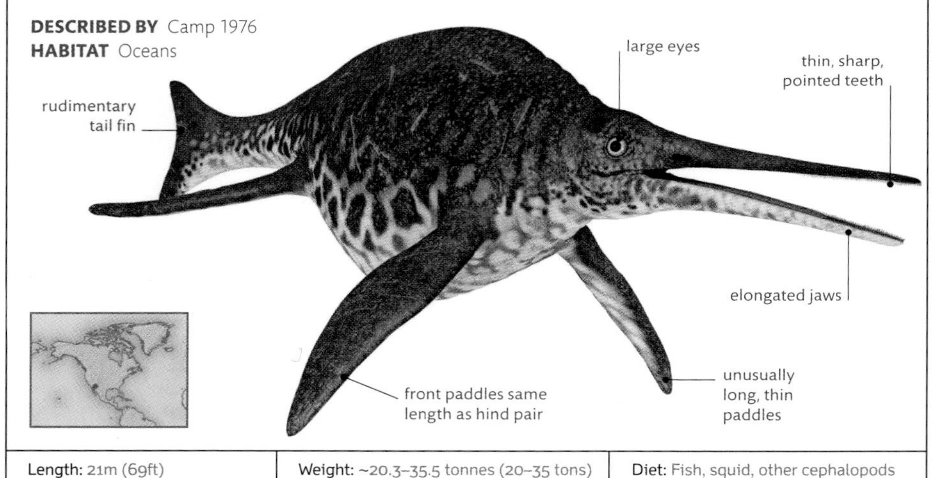

Length: 21m (69ft)	Weight: ~20.3–35.5 tonnes (20–35 tons)	Diet: Fish, squid, other cephalopods

Group: DIAPSIDA	Subgroup: Tanystropheidae	Time: 230mya

Tanystropheus

TAN-EE-STRO-FEE-US

This reptile is one of the strangest to have ever existed. Its incredibly long neck was composed of only 13 vertebrae, which were so elongated that they were first thought to be leg bones. The neck length has led to much speculation about this animal's lifestyle, as it was not well-adapted for either walking or swimming. Current opinion is that it used its neck to fish from sea or lake shores or from shallow water.

DESCRIBED BY von Meyer 1852
HABITAT Shorelines

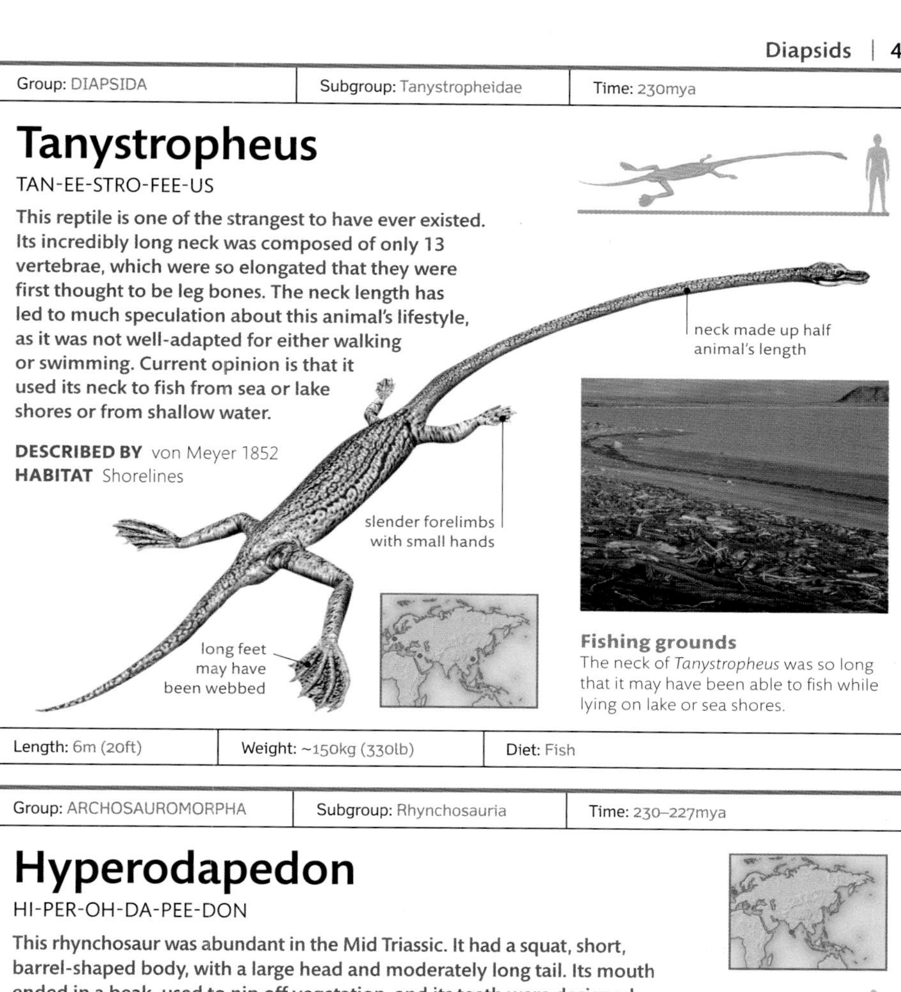

Fishing grounds
The neck of *Tanystropheus* was so long that it may have been able to fish while lying on lake or sea shores.

Length: 6m (20ft)	Weight: ~150kg (330lb)	Diet: Fish

Group: ARCHOSAUROMORPHA	Subgroup: Rhynchosauria	Time: 230–227mya

Hyperodapedon

HI-PER-OH-DA-PEE-DON

This rhynchosaur was abundant in the Mid Triassic. It had a squat, short, barrel-shaped body, with a large head and moderately long tail. Its mouth ended in a beak, used to nip off vegetation, and its teeth were designed for efficient chopping of tough plant material. It had several rows of teeth on the upper jaw. A groove ran through the middle row of teeth and, on the lower jaw, a single row of teeth fitted into it when the mouth was closed.

DESCRIBED BY
Huxley 1859
HABITAT
Woodland

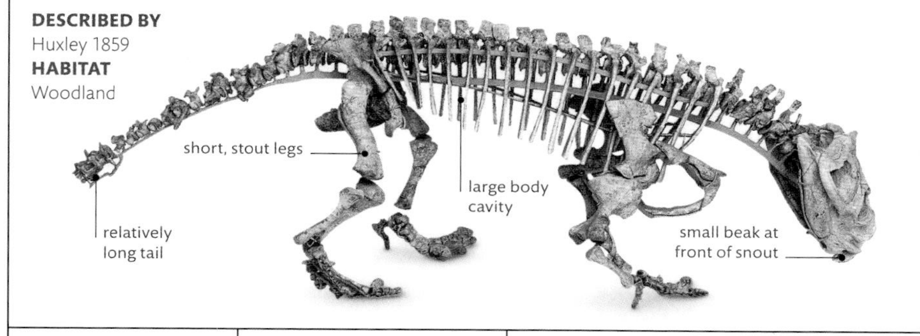

Length: 1.2m (4ft)	Weight: ~40kg (88lb)	Diet: Seed ferns

Group: ARCHOSAUROMORPHA	Subgroup: Archosauriformes	Time: 252–247mya

Proterosuchus

PRO-TEH-RO-SU-KUSS

***Proterosuchus* was one of the earliest known archosaurs. It had a large, heavy body, and its legs were angled out from the body, resulting in a sprawling, lizard-like gait. It probably lived a lifestyle similar to that of crocodiles, spending most of its time hunting in rivers. Its teeth were sharp and curved backwards, and there were also primitive teeth on the palate. *Proterosuchus* had a pointed, downwardly-hooked snout.**

DESCRIBED BY
Broom 1903
HABITAT Riverbanks

large head with long snout

long, heavy tail

sprawling, stout limbs

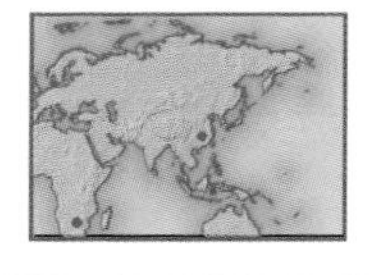

Length: 2m (6½ft)	Weight: Not calculated	Diet: Fish, herbivorous dicynodonts

Group: ARCHOSAURIFORMES	Subgroup: Pseudosuchia	Time: 222–212mya

Parasuchus

PAR-A-SOOK-US

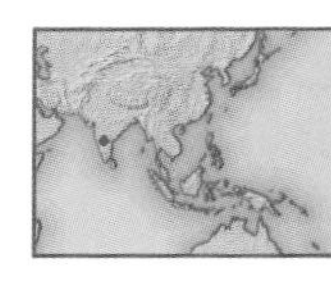

***Parasuchus*, whose name means "near crocodile", looked remarkably like a modern crocodile. Its throat and back were protected by heavy armoured plates (or scutes), and the belly was strengthened by a dense arrangement of abdominal ribs. The skull was long, with a slender snout. The jaws were lined with conical teeth. Nostrils positioned on top of the head allowed *Parasuchus* to breathe while underwater.**

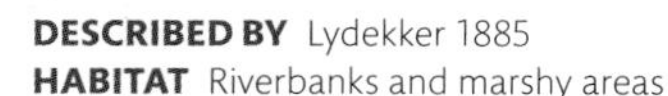

DESCRIBED BY Lydekker 1885
HABITAT Riverbanks and marshy areas

back protected by heavy scutes

nostrils on top of snout

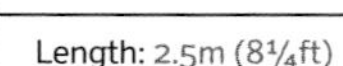

Length: 2.5m (8¼ft)	Weight: Not calculated	Diet: Fish

Group: ARCHOSAURIFORMES	Subgroup: Pseudosuchia	Time: 235–206mya

Stagonolepis

STAG-ON-OH-LEP-IS

Like other members of its family, *Stagonolepis* was herbivorous and had a deep, low-slung body adapted to accommodate the longer intestines needed for digesting plants. It was heavily armoured, with rectangular plates covering the whole of the back and tail. A row of short spikes ran along each side of the animal, and the underside of the belly and tail was covered with more bony plates.

DESCRIBED BY Agassiz 1844
HABITAT Forests

Length: 2.5m (8¼ft)	Weight: ~200kg (440lb)	Diet: Plants

Group: ARCHOSAURIFORMES	Subgroup: Pseudosuchia	Time: 228–200mya

Desmatosuchus

DES-MAT-OH-SOOK-US

***Desmatosuchus* ("link crocodile") was a particularly heavily armoured animal, which superficially resembled a short-snouted crocodile. Rows of rectangular plates covered its back and tail, and a row of short spikes ran along the flanks. Long spikes jutted out from each shoulder. The underside of the belly was also covered in bony plates. The body was long and deep, with relatively short legs. The cheek teeth were weak and peg-like.**

DESCRIBED BY Case 1920
HABITAT Forests

toothless snout

rectangular bony plates along back and tail

shoulder spikes up to 45cm (18in) long

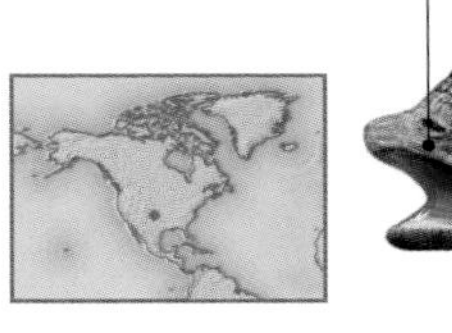

Length: 5m (16ft)	Weight: ~300kg (660lb)	Diet: Plants

Group: ARCHOSAUROMORPHA	Subgroup: Archosauriformes	Time: 235mya

Euparkeria

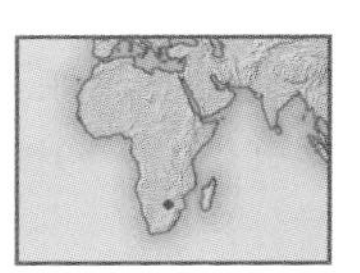

EWE-PARK-ER-EE-A

Euparkeria was an early archosaur, and was unusual among its contemporaries in that the relative length of its hind legs to its forelegs was greater than that of contemporary reptiles. It probably spent most of its time on all fours, but was capable of rising onto its hind legs to run. Its tail made up about half its body weight, and was held outstretched for balance as it ran. The body was small and slim, with thin bony plates covering the middle of the back and tail. The head was large but light, and the many teeth were sharp, serrated, and curved backwards, indicating a carnivorous way of life.

DESCRIBED BY Broom 1913
HABITAT Woodland

ridge of light bony plates running along back

lightly built head despite large size

four-fingered hands

Length: 60cm (24in)	Weight: ~13.5kg (30lb)	Diet: Meat

Group: ARCHOSAURIA	Subgroup: Pterosauria	Time: 227–208mya

Peteinosaurus

PET-INE-OH-SAW-RUS

Peteinosaurus ("winged lizard") is one of the earliest vertebrates to show evidence of active, flapping flight. It had a light-boned skeleton, like that of a modern bird. Its long tail was stiffened with bony ligaments to stabilize it during flight. It had cone-shaped teeth for crunching insects.

DESCRIBED BY Wild 1978
HABITAT Swamps and river valleys

wing membrane stretched from fourth finger

large, light head

long, bony tail

Length: 60cm (24in)	Weight: Not calculated	Diet: Flying insects

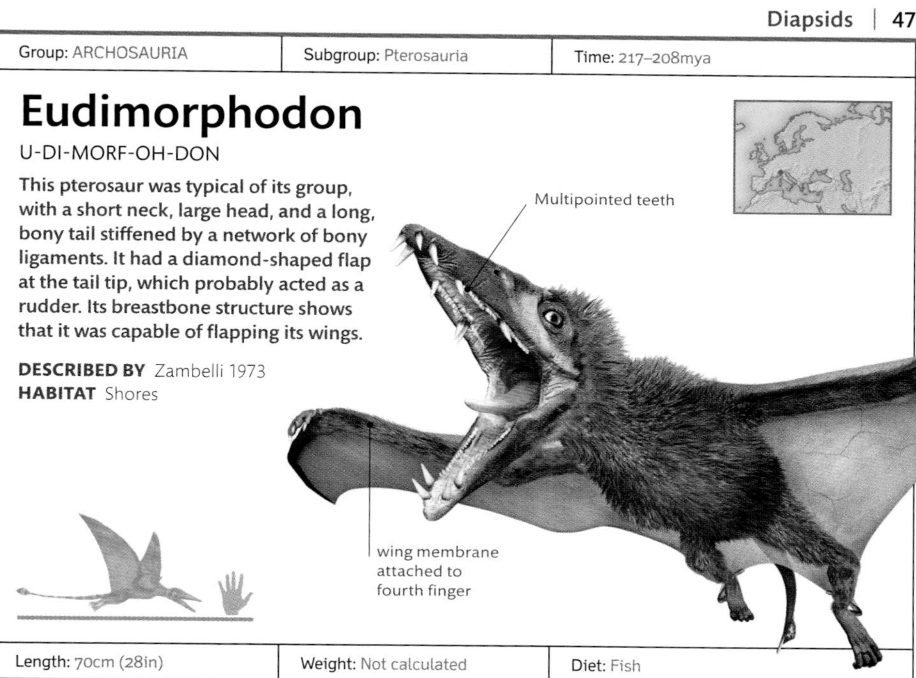

Group: ARCHOSAURIA	Subgroup: Pterosauria	Time: 217–208mya

Eudimorphodon

U-DI-MORF-OH-DON

This pterosaur was typical of its group, with a short neck, large head, and a long, bony tail stiffened by a network of bony ligaments. It had a diamond-shaped flap at the tail tip, which probably acted as a rudder. Its breastbone structure shows that it was capable of flapping its wings.

DESCRIBED BY Zambelli 1973
HABITAT Shores

Length: 70cm (28in)	Weight: Not calculated	Diet: Fish

Group: ARCHOSAURIA	Subgroup: Dinosauromorpha	Time: 230mya

Lagosuchus

LAY-GO-SOOK-US

***Lagosuchus*, or "rabbit crocodile", belonged to the group that gave rise to the dinosaurs, although it is unlikely that it was their direct ancestor. It was indeed strikingly similar in appearance to small, early theropod dinosaurs. Its body was slim and lightly built, the tail was long and flexible, and the hind legs were long and thin, with shin bones much longer than the thighs. *Lagosuchus* ran on its hind legs.**

long, slim snout

thighs shorter than shins

long feet

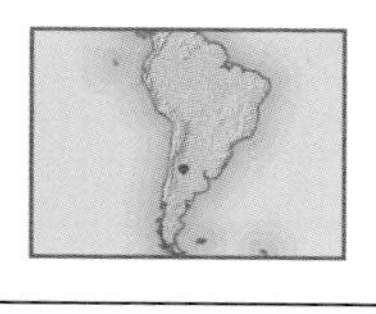

DESCRIBED BY
Romer 1971
HABITAT Forest

Length: 30cm (12in)	Weight: ~90g (3¼oz)	Diet: Meat

Group: SAURISCHIA	Subgroup: Herrerasauridae	Time: 233mya

Gnathovorax

NATH-OH-VOR-ACKS

Gnathovorax ("devouring jaw") is known from the most complete skeleton of any herrerasaurid. An analysis of its teeth shows that it was a carnivore, and its diet may have consisted of rhynchosaurs and cynodonts, which were found with the skeleton in Brazil. Its brain was still primitive, but the regions that controlled the eyes and head were enlarged, probably for its predatory lifestyle.

DESCRIBED BY Pacheco et al. 2019
HABITAT Semi-arid coastal plains

Length: 2m (6½ft)	Weight: Not calculated	Diet: Meat

Group: SAURISCHIA	Subgroup: Herrerasauridae	Time: 231–225 mya

Staurikosaurus

STORE-EE-KOH-SAW-RUS

Staurikosaurus ("Southern Cross lizard") was a primitive, bipedal dinosaur. It had the typical theropod body shape -long, slim tail, long, powerful hind legs, and short arms. The back was held horizontally, with the tail used for balance. The lower jaw had a joint that allowed the tooth-bearing part to move independently of the back part of the jaw.

DESCRIBED BY Colbert 1970
HABITAT Forests and scrub

Length: 2m (6½ft)	Weight: ~20–40kg (44–88lbs)	Diet: Meat

Group: SAURISCHIA	Subgroup: Herrerasauridae	Time: 228mya

Herrerasaurus

HERR-RAY-RAH-SAW-RUS

Analysis of a complete skeleton of *Herrerasaurus* ("Herrera's lizard") discovered in the 1990s, confirmed that it was one of the most primitive dinosaurs. However, palaeontologists are still disputing whether it can be classified as a theropod. *Herrerasaurus* was an agile hunter, probably able to prey on slower creatures such as *Hyperodapedon* (see p.43). It had powerful hindlimbs, and a long tail used for balance. It probably ran with its back nearly horizontal to balance the body at the hips. The jaws were lined with sharp, long teeth that were curved backwards. New evidence suggests that the skull was more boxy than tapered at the snout.

DESCRIBED BY Reig 1963
HABITAT Woodland

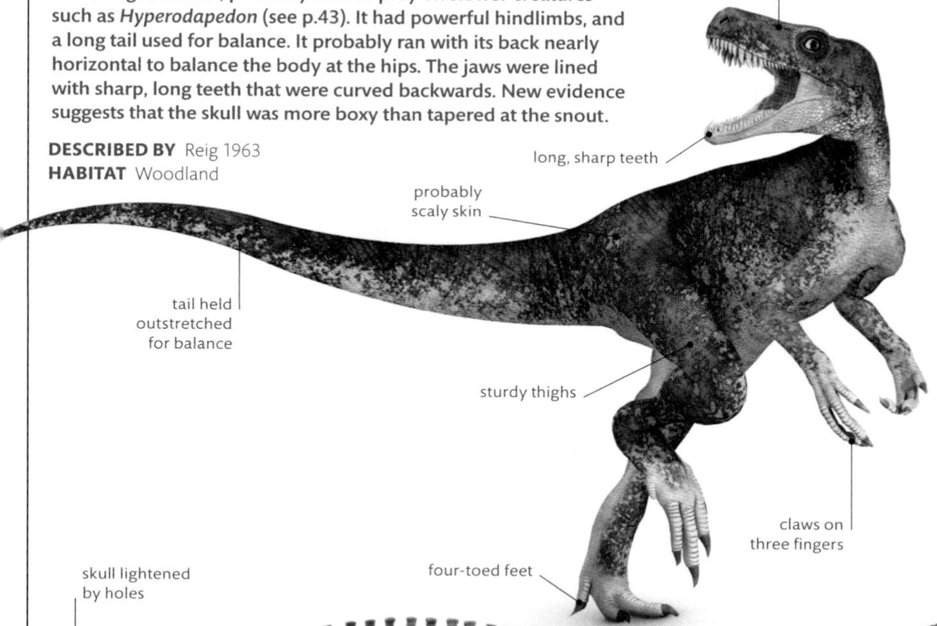

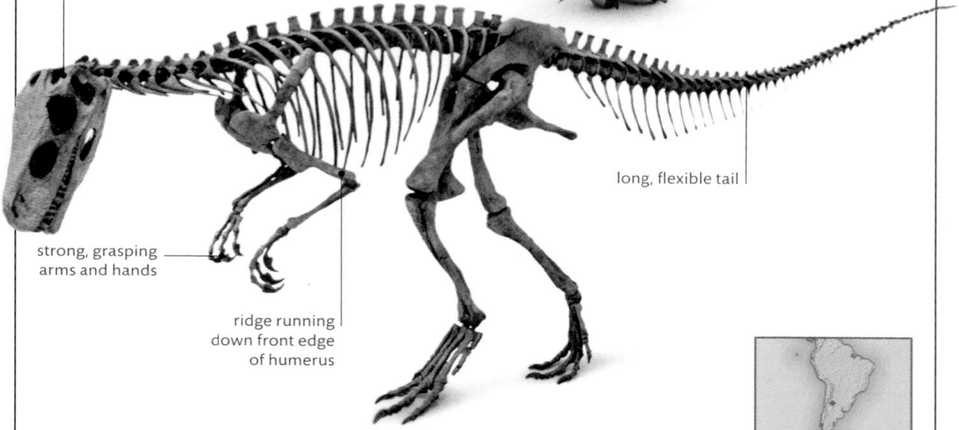

Skeletal reconstruction

Length: 3.1m (10ft)	Weight: ~210kg (460lb)	Diet: Meat

Group: SAURISCHIA	Subgroup: Sauropodomorpha	Time: 210mya

Efraasia

EE-FRAA-ZIA

This early sauropodomorph was named after its discoverer, E. Fraas. It was lightly built, with a small head, fairly long neck, and long tail. Its legs were longer than its arms, and its five-fingered hands had a large thumb claw. It may have walked on all fours to browse, rising onto its hind legs to run.

DESCRIBED BY Galton 1973
HABITAT Dry upland plains

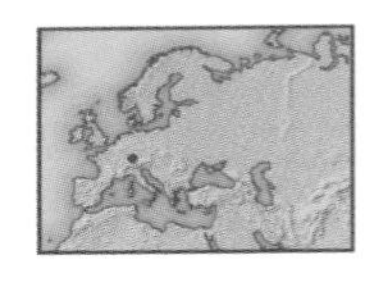

Length: 6.5m (21ft)	Weight: ~400kg (880lb)	Diet: Plants, possibly some meat

Group: SAURISCHIA	Subgroup: Sauropodomorpha	Time: 203–201mya

Thecodontosaurus

THEE-COH-DONT-OH-SAW-RUS

This dinosaur was one of the earliest dinosaurs to be discovered, but it was not recognized as a dinosaur until later. *Thecodontosaurus* is one of the most primitive sauropodomorphs known. Its name ("socket-toothed lizard") was given because its saw-edged teeth were embedded in the sockets in the jaw bones. *Thecodontosaurus* had a relatively small head and a long tail. Mainly bipedal, it seems to have had some ability to walk on all fours.

DESCRIBED BY Riley and Stutchbury 1836
HABITAT Desert plains, dry upland areas

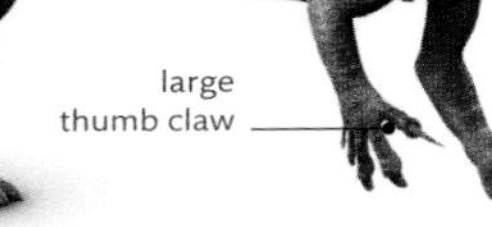

Length: 2.1m (7ft)	Weight: ~50kg (110lb)	Diet: Plants, possibly omnivorous

Group: SAURISCHIA	Subgroup: Sauropodomorpha	Time: 214–204mya

Plateosaurus

PLAT-EE-OH-SAW-RUS

This "broad lizard" was one of the most common dinosaurs of the Late Triassic Period. The large number of fossil finds suggest that *Plateosaurus* may have lived in herds and migrated to avoid seasonal droughts. It walked on its hind legs and gathered food from high vegetation. The thumbs on its hands were partially opposable, allowing it to grasp food. Its small skull was deeper than in most other sauropodomorphs, and its fairly short neck was thinner. *Plateosaurus* had many small, leaf-shaped teeth, and the hinge of the lower jaw was low-slung to give greater leverage. These factors indicate a diet primarily composed of plants. This dinosaur also had a very large nasal chamber, but the reason for this is as yet unknown.

DESCRIBED BY Meyer 1837
HABITAT Dry plains, desert

low hinge on lower jaw
long finger and hand bones
strong thigh bone
flexible tail about half dinosaur's length

Skeletal reconstruction

beak-like upper jaw
fairly short, thin, flexible neck
bulky body
hindlimbs longer than forelimbs
clawed fingers
three forward-facing clawed toes

Length: 8m (26ft)	Weight: ~4 tonnes (4.4 tons)	Diet: Leaves, small amounts of meat

Group: SAURISCHIA	Subgroup: Sauropodomorpha	Time: 228mya

Eoraptor

EE-OH-RAP-TOR

Eoraptor ("dawn raptor") is regarded as one of the earliest dinosaurs. It was a very small, lightly-built, bipedal carnivore with hollow bones. The head was long and slim, with many small, sharp teeth. The arms were far shorter than the legs, and had five-fingered hands, although two of the fingers were reduced. *Eoraptor* appears to lack the specialized features of any of the major groups of dinosaurs, and for this reason is consistent with what would be expected for the earliest dinosaur ancestors.

DESCRIBED BY Sereno et al. 1993
HABITAT Forests

Length: 1m (39in)	Weight: ~10kg (22lbs)	Diet: Meat

Group: SAURISCHIA	Subgroup: Theropoda	Time: 213–190mya

Coelophysis

SEE-LOW-FIE-SIS

Coelophysis ("hollow face") was a small, lightly built, early dinosaur with open skull bones (hence its name). The body was long and slim, and the head was pointed, with many small, serrated teeth. *Coelophysis* is one of the best-known dinosaurs due to the excavation of dozens of skeletons from New Mexico, USA (see right). Many of these fossils contained the remains of their prey. It was suggested that these were bones of young *Coelophysis*, and was seen as evidence of cannibalism. However, the bones are now interpreted to be those of some kind of crocodylomorph.

DESCRIBED BY
Cope 1889
HABITAT
Desert plains

Length: 2m (6½ft)	Weight: ~12kg (26lb)	Diet: Meat, perhaps including birds, carrion

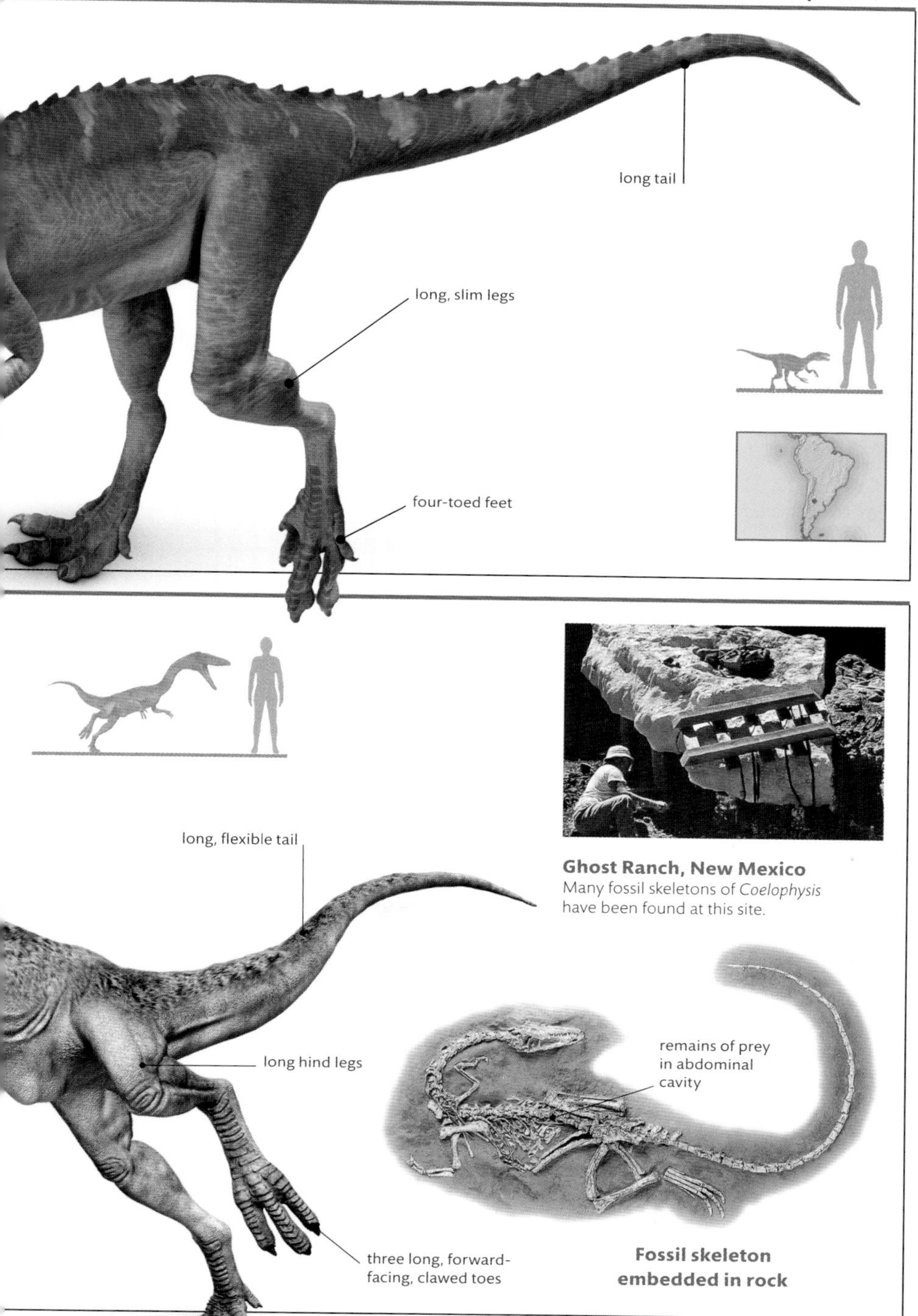

Ghost Ranch, New Mexico
Many fossil skeletons of *Coelophysis* have been found at this site.

Fossil skeleton embedded in rock

SYNAPSIDS

SYNAPSIDS WERE HIT particularly hard in the end-Permian extinction. Therapsids, which had replaced pelycosaurs as the dominant synapsids in the Permian, were represented only by a few survivors. These were mostly small, rapidly-breeding burrowers that were able to quickly adapt to the harsh Early Triassic landscape.

Gorgonopsians, the Permian top predators, went extinct in the end-Permian extinction. Dicynodonts survived, and in the Early Triassic, they were incredibly abundant throughout the world. Alongside the diapsid rhynchosaurs, dicynodonts became important large herbivores, with some, such as *Lisowicia*, reaching the size of elephants. Despite their success, they went extinct at the end of the Triassic.

Another surviving therapsid lineage, the cynodonts, eventually occupied a wide range of ecological niches - including large carnivores and herbivores - but it was the smaller members of this group that would be the most successful. These nocturnal insectivores became highly specialized, developing complex teeth, keen hearing, and coats of fur for insulation. These changes were linked to a shift in metabolism, and these small cynodonts started to produce and regulate their own body heat. This new lifestyle required huge amounts of energy to sustain, but it helped them stay active regardless of environmental conditions. Cynodonts were the only synapsid survivors of the Triassic, and their descendants, the mammals, still thrive today.

Group: THERAPSIDA	Subgroup: Dicynodontia	Time: 247–242mya

Sinokannemeyeria

SY-NOH-KAN-EH-MEY-REE-A

This long-snouted member of the dicynodont ("two-dog teeth") group of synapsids was well adapted for a lifestyle as a terrestrial herbivore. Its head was massive, with large openings for the eyes, nostrils, and jaw muscles. These reduced the weight of the skull. A hinge between the lower jaw and the skull allowed the jaws to move backwards and forwards with a shearing action. Although the jaws were toothless, this motion would have ground up the toughest vegetation. The front of the jaw had a small horn-covered beak, and there were two small tusks growing from bulbous projections on the upper jaw. These tusks could have been used to dig up roots. *Sinokannemeyeria* had relatively short, stumpy legs, which were held in a slightly sprawling gait to the sides of its body. The limb girdles were formed into large, heavy plates of bone to support the weight of the wide, heavily built body. *Sinokannemeyeria* was probably not a fast or agile mover.

DESCRIBED BY Young 1937
HABITAT Plains and woodland

large attachment points for jaw muscles

large eye sockets

turtle-like horny beak at front of mouth

broad toes on flat feet

Length: 1.8m (6ft)	Weight: ~100kg (220lb)	Diet: Tough vegetation, roots

Group: THERAPSIDA	Subgroup: Dicynodontia	Time: 255–250mya

Lystrosaurus

LIS-TRO-SAW-RUS

Lystrosaurus, whose name means "shovel lizard", was a heavily built, early dicynodont with a short, stubby tail. It had two tusk-like fangs made of horn. It used to be thought of as a type of reptilian hippopotamus because of its long, downturned snout and the placement of its nasal openings high on the snout. However, new analysis suggests a terrestrial lifestyle: the skull and jaw features were adaptations to a fibrous diet, the pelvis was well-developed, and the hind limbs were semi-erect.

long spine supported barrel-shaped body

long snout with two horn fangs

short, semi-erect limbs

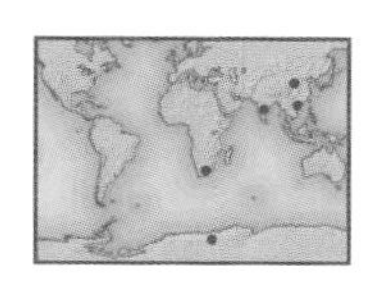

DESCRIBED BY Cope 1870
HABITAT Dry floodplains

Length: 1m (3¼ft)	Weight: ~92kg (200lbs)	Diet: Plants

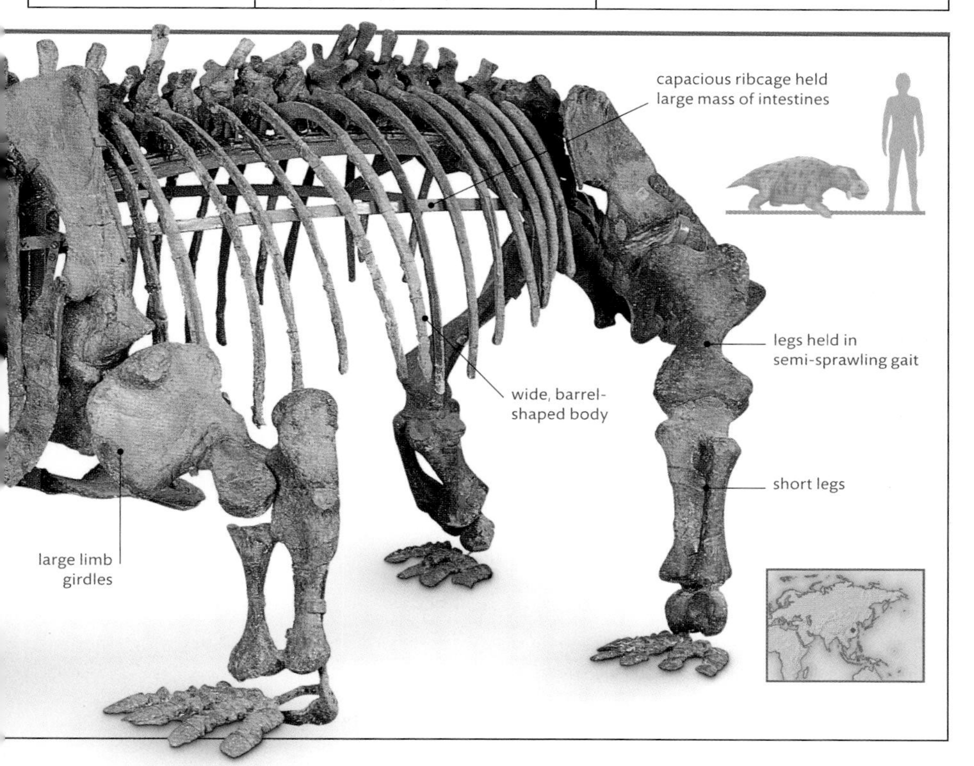

Group: THERAPSIDA	Subgroup: Dicynodontia	Time: 220–216mya

Placerias

PLAH-SEE-REE-AS

Placerias was a heavily built powerful animal, with the characteristic dicynodont jaw structure. Its jaw margins were toothless, except for two horny tusks near the front of the mouth. The tip of the snout ended in a beak used for uprooting plants. *Placerias* was one of the last of the dicynodont species.

DESCRIBED BY Lucas 1904
HABITAT Flood plains

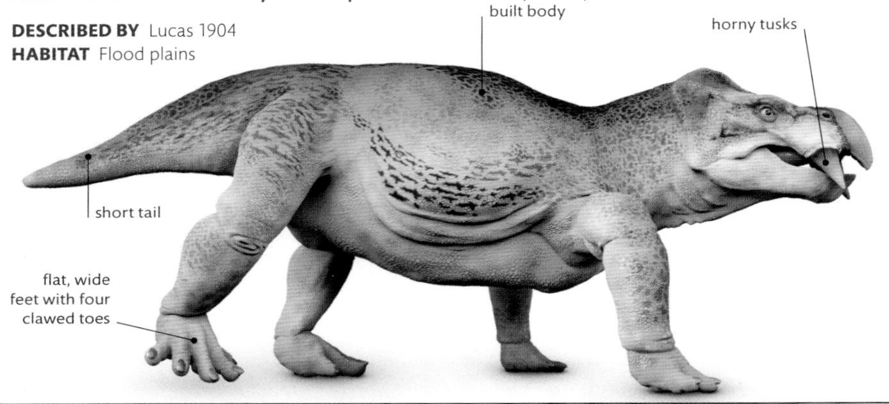

Length: 1.2–3.5m (4–11ft)	Weight: ~Up to 1 tonne (1 ton)	Diet: Plants

Group: THERAPSIDA	Subgroup: Cynodontia	Time: 247–237mya

Cynognathus

SY-NOG-NAY-THUS

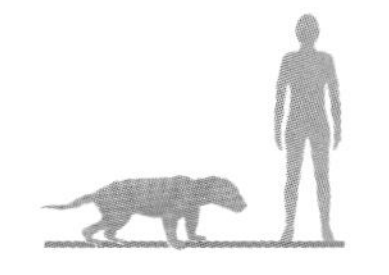

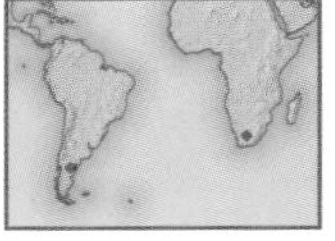

This wolf-sized carnivore was a ferocious Triassic predator. It was one of the largest cynodonts ("dog-toothed" synapsids), and was heavily built, with strong legs positioned directly under its body, and a tail shorter than that of most other reptiles. zzzIts head was more than 30cm (12in) long. It is thought that it was probably endothermic, its skin may have been covered with hair, and it probably laid eggs. *Cynognathus* ("dog jaw") had dog-like teeth, with cutting incisors, long canines, and shearing set of cheek teeth. The jaws had a wide gape and a very powerful bite.

DESCRIBED BY Seeley 1895
HABITAT Woodland

Fossil skull

Length: 1.5m (5ft)	Weight: ~40–50kg (88–110lb)	Diet: Herbivorous therapsids

Group: THERAPSIDA | Subgroup: Cynodontia | Time: 251–247mya

Thrinaxodon

THRIN-AXE-OH-DON

This small, solidly built carnivore had a long body, which was distinctly divided into thoracic and lumbar regions (the first time this was seen among vertebrates). The division was marked by the ribs, which were borne only on the thoracic vertebrae. This suggests that *Thrinaxodon* ("trident tooth") probably had a diaphragm, as modern mammals do. One of the foot bones had evolved into a heel, so that the foot could be levered clear of the ground for more efficient running. The teeth were set into a single bone, which made the jaws stronger.

DESCRIBED BY Seeley 1894
HABITAT Woodland

body may have been covered with hair

teeth with three sharp cusps

toes of equal length

Length: 50cm (20in) | Weight: Not calculated | Diet: Meat

Group: CYNODONTIA | Subgroup: Mammaliaformes | Time: 208–174mya

Eozostrodon

EE-OH-ZOH-STROH-DON

This shrew-like animal was one of the earliest true mammals. It probably laid eggs, but is thought to have fed its young with milk produced by mammary glands. Its four short legs had a slightly sprawling gait and ended in five-toed feet with claws. Its tail was long and possibly hairy. Its snout was long, slim, and contained true mammalian teeth (the cheek teeth comprised simple premolars and molars with multiple sharp cusps, and were replaced only once during the animal's lifetime). The large eyes suggest that *Eozostrodon* was probably a nocturnal hunter, and the sharp teeth indicate a diet predominantly of insects and small animals. Recent analysis suggests that this genus is distinct from the later-named *Morganucodon*.

multi-cusped molars

Jaw bone

DESCRIBED BY Parrington 1941
HABITAT Forest floors

small body covered with dense hair

slim pointed snout may have had whiskers

large eyes

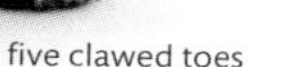

Length: 10cm (4in) | Weight: ~150g (5oz) | Diet: Insects, small animals

JURASSIC PERIOD

201–145 MILLION YEARS AGO

AT THE BEGINNING of the Jurassic Period, the world was still hot and arid. Major continental movements caused climate change throughout the period: increased rainfall led to a reduction in desert areas, climates were more humid, and, as the period progressed, habitats in general became more green and luxuriant. However, lake deposits show that there were millions of cycles of dry and wet periods. Most Jurassic plants were still mainly primitive types of gymnosperms. There were also ferns, horsetails, and club mosses, and huge forests of these plants spread worldwide. Continental movements led to the creation of warm, shallow seas, which became filled with coral reefs and new forms of marine life, including large reptiles. Pterosaurs were present in large numbers, and the earliest birds appeared at the end of the period.

JURASSIC LIFE

Early, bipedal sauropodomorphs still dominated Early Jurassic habitats, but as the period progressed, they were usurped by the giant sauropods, such as *Apatosaurus*. The first armoured dinosaurs – the stegosaurs – also appeared. Protective armour may have been an evolutionary response to the emergence of ever-larger carnivorous dinosaurs. The first true mammals evolved, and probably survived by remaining small and living a mainly nocturnal existence.

relatively short tail

Marine predators
Giant pliosaurs such as *Liopleurodon* (above), filled the warm Jurassic seas, hunting large prey including other marine reptiles. They had developed an efficient "flying" action for swimming rapidly through the water.

wing supported by greatly elongated fourth finger

Later pterosaurs
In the Late Jurassic, new forms of pterosaurs, such as *Anurognathus* (above) appeared. Their main feature was a very short tail that gave them greater agility in the air.

4,600mya	4,000mya	3,000mya

JURASSIC LANDMASSES

At the start of the Jurassic Period, there was once again a supercontinent called Gondwana south of the Equator. Later in the period (see map), this split to form the landmasses that were to become present-day Australia, Antarctica, India, Africa, and South America. Laurasia (comprising the future North America and Europe) had begun to take shape in the northern hemisphere.

DIVERSITY

In the early Jurassic Period, the maximum size of dinosaurs increased, and this trend continued throughout the period. Because of this, mammals (such as *Sinoconodon*, below) and lizards, tended to be rather small in order to maintain their ecological niches. In the seas, the ichthyosaurs, a fish-like group of marine reptiles, were at their zenith, sharing the warm Jurassic oceans with other marine predators, including plesiosaurs, sharks, and marine crocodiles.

Trees
Many forests of evergreen trees spread throughout the world. *Araucaria* were conifers. A fossilized cone is pictured, left.

THEROPODS

THE JURASSIC saw the true transition to dinosaur-dominated ecosystems, and theropods replaced pseudosuchians as top predators. Some of the early lineages, like dilophosaurids, became slightly larger, but it was the ceratosaurs and tetanurans that dominated the food chain. The ceratosaur lineage spawned the abelisaurs, which remained top predators in the southern hemisphere until the end-Cretaceous extinction. Tetanurans diverged in skull shape, with some groups developing elaborate crests and horns, and others an elongated snout – taken to an extreme in the bizarre spinosaurs.

As these lineages were getting bigger, other groups like the coelurosaurs were getting smaller, so theropods occupied a wide variety of ecological roles. Many of the lineages that would come to rule the Cretaceous originated in the Jurassic, including dromaeosaurs, tyrannosaurs, and troodontids. Some of these coelurosaurs, which were fast runners with relatively long arms, started to use their feathered forelimbs in unique ways, experimenting with generating lift by flapping or gliding. It is now clear that active powered flight originated independently multiple times in these dinosaurs, but one particular dinosaur, *Archaeopteryx* (see pp.70–71), was especially adept at flying. Its relatives continue to rule the skies today.

Group: THEROPODA	Subgroup: Ceratosauria	Time: 156–149mya

Ceratosaurus

SER-AT-OH-SAW-RUS

Ceratosaurus ("horned lizard") was named for the short horn above its nose. Another striking feature was the line of bony plates that ran down its back – so far, it is the only theropod known to have had them. It had strong, yet short, arms, with four fingers on each hand. Three of the fingers were clawed. It had a deep, broad tail, and feet with three large toes and a reduced back toe. Its teeth were long and blade-like. Although it superficially resembled a carnosaur such as *Allosaurus*, it was more primitive, and its tail was flexible rather than stiffened with bony ligaments as was the case with the carnosaurs.

DESCRIBED BY Marsh 1884
HABITAT Forested plains

Length: 4.5–6m (15–20ft)	Weight: ~1 tonne (1 ton)	Diet: Herbivorous dinosaurs, other reptiles

Group: THEROPODA	Subgroup: Dilophosauridae	Time: 199–182mya

Dilophosaurus

DI-LOAF-OH-SAW-RUS

Dilophosaurus ("two-ridged lizard") was named for the striking pair of bony crests that adorned its head. These were so thin and fragile that they were almost certainly only used for sexual display. *Dilophosaurus* had a more primitive body structure than that of coelurosaurs and carnosaurs – it had a large head, but was lightly built, with a slender neck, body, and tail. Once thought to have been closely related to *Coelophysis* for its four-fingered hand and a notch in its upper jaw, it is now more likely to be somewhat intermediate between *Coelophysis* and the larger Ceratosaurs.

DESCRIBED BY Welles 1970
HABITAT Riverbanks

bony, semicircular crests

three long, forward-facing, clawed toes

flexible tail

long, slim, powerful hind legs

Length: 6m (20ft)	Weight: ~500kg (½ ton)	Diet: Small animals, perhaps fish or carrion

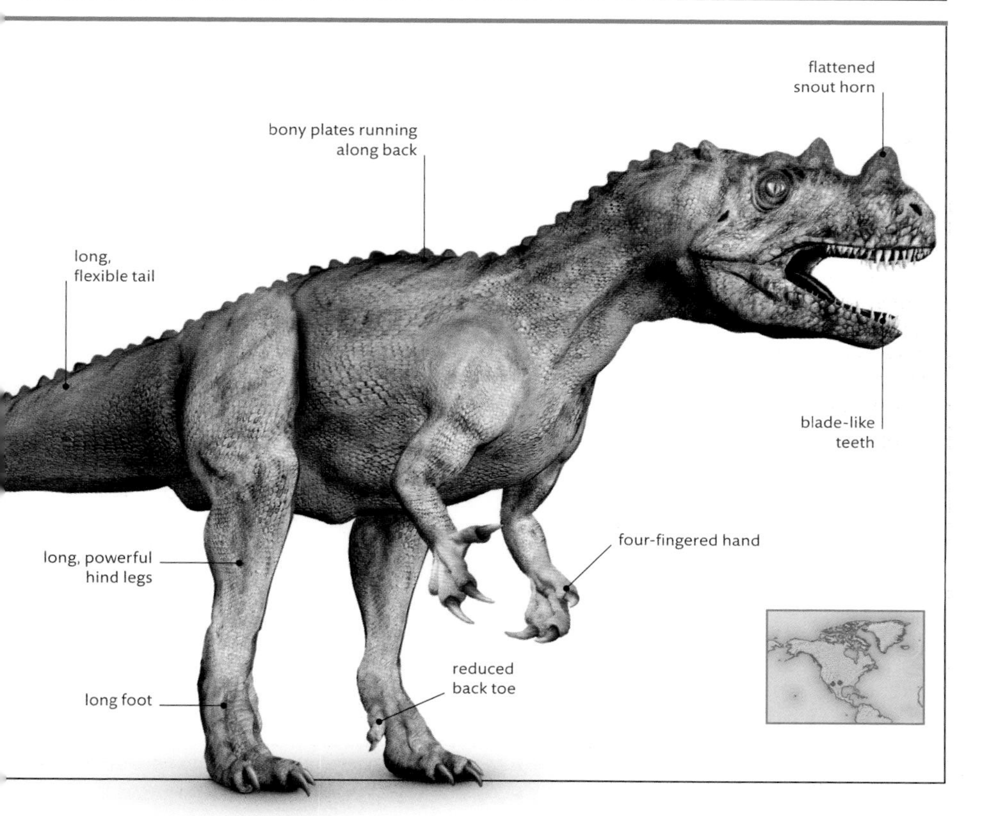

Group: THEROPODA	Subgroup: Tetanurae	Time: 171–161mya

Gasosaurus

GAS-OH-SAW-RUS

This theropod was given its unusual name ("gas lizard") in honour of the Chinese gas-mining company that was working in the quarry where the fossil remains were excavated. The single species *Gasosaurus constructus* is represented by a single specimen. Remains included a humerus, the pelvis, and a femur. Little is known about *Gasosaurus*, and its classification is still uncertain. It may be a primitive carnosaur; however, some features of the leg bones suggest that it may be an early coelurosaur instead. If so, it would be one of the oldest coelurosaurs so far discovered. *Gasosaurus* had the typical theropod body shape of a large head, long, powerful legs ending in three forward-facing clawed toes, and a long, stiff tail. It had very short arms, but they were longer than those of later carnosaurs.

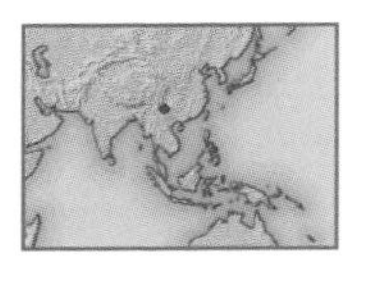

DESCRIBED BY
Dong and Tang 1985
HABITAT Woodland

Length: 3.1m (10ft)	Weight: ~150kg (330lb)	Diet: Large herbivorous dinosaurs

Group: CARNOSAURIA	Subgroup: Allosauroidea	Time: 166–163mya

Piatnitzkysaurus

PEAT-NYITS-KEE-SAW-RUS

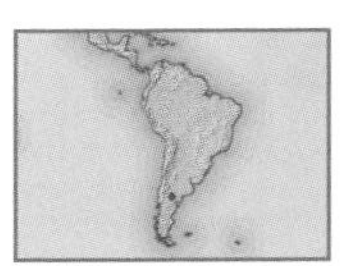

Piatnitzkysaurus, whose name means "Piatnitzky lizard", seems to have been a tetanuran (stiff-tailed) theropod. Two partial skeletons have been excavated to date, and its exact classification is still under debate by palaeontologists. It had a very similar body to *Allosaurus* (see p.64), but its arms were longer. Its head was large, and was carried on a short, muscular neck. The arms were relatively small. The body was bulky, and the tail was long and stiff.

DESCRIBED BY
Bonaparte 1979
HABITAT Woodland

Length: 4.3m (14ft)	Weight: ~400kg (880lb)	Diet: Herbivorous dinosaurs

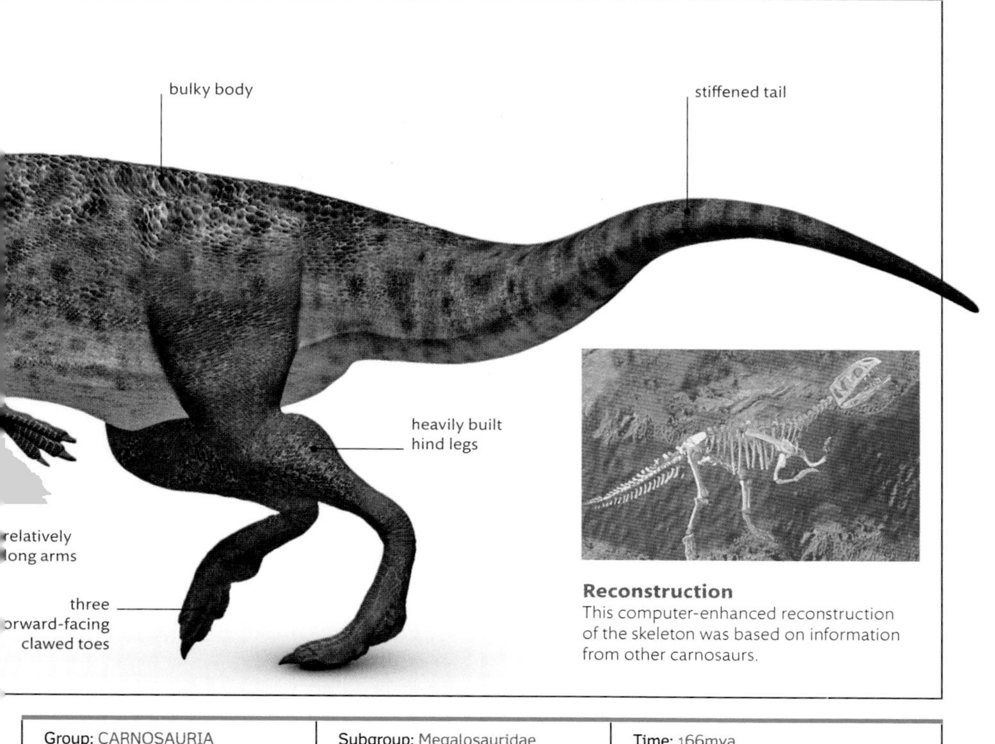

Reconstruction
This computer-enhanced reconstruction of the skeleton was based on information from other carnosaurs.

Group: CARNOSAURIA	Subgroup: Megalosauridae	Time: 166mya

Megalosaurus

MEG-A-LOH-SAW-RUS

The first dinosaur to be scientifically named and identified, *Megalosaurus* ("great lizard") was a large, bulky, carnivorous theropod. It had a massive head carried on a short, muscular neck. Its arms were short, but strong. Fossil trackways found in southern England show how *Megalosaurus* walked with its toes pointing slightly inwards, with its tail probably swinging from side to side for balance.

DESCRIBED BY Buckland 1824
HABITAT Forests

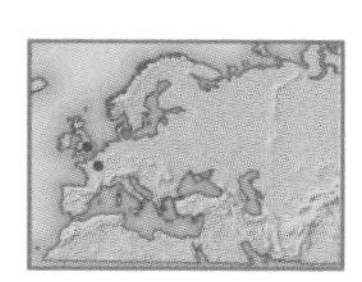

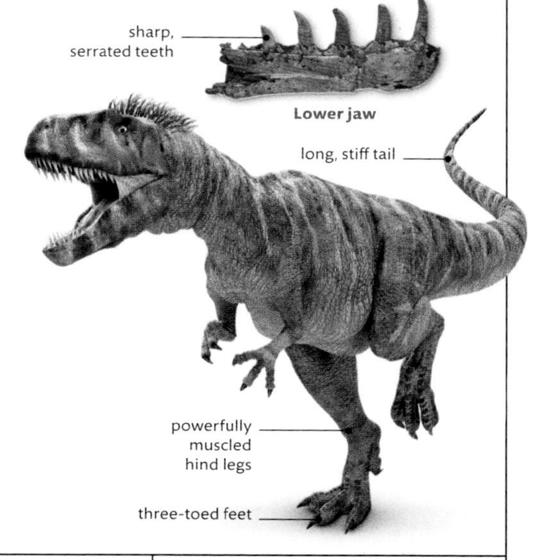

Length: 9m (30ft)	Weight: ~1–2 tonnes (1–2 tons)	Diet: Large herbivorous dinosaurs

Group: CARNOSAURIA	Subgroup: Allosauroidea	Time: 156–149mya

Allosaurus

AL-OH-SAW-RUS

Allosaurus **("different lizard") was the most abundant, and probably the largest, predator in the late Jurassic period in the lands that were to become North America. It was typical of its family of meat-eating dinosaurs, with a massive head, short neck, and bulky body. Its tail was long and deep, with a thin, stiff end. Its three-fingered forelimbs were strong, with large claws.** ***Allosaurus*** **had distinctive bony bumps over the eyes, and a narrow ridge of bone running between them down to the tip of the snout. These features may have been used for display. Although the head was massive, it was lightened by several large openings (fenestrae) between the bones. Expandable joints between the skull bones allowed the jaws to gape open sideways to swallow large mouthfuls. Despite its large size and lack of speed, some palaeontologists believe that** ***Allosaurus*** **was agile enough to fell the giant plant-eating dinosaurs of the time.**

DESCRIBED BY
Marsh 1877
HABITAT Plains

distinctive ridge of bone along snout

teeth with serrated front and back edges

FOSSIL FINDS

Many complete skeletons of *Allosaurus* have been excavated. Many partial skeletons have been found in the Cleveland-Lloyd Dinosaur Quarry in Utah. This site is thought to have been a place where herbivorous dinosaurs became stuck in mud and attracted predators, which also became trapped as they attacked the helpless prey.

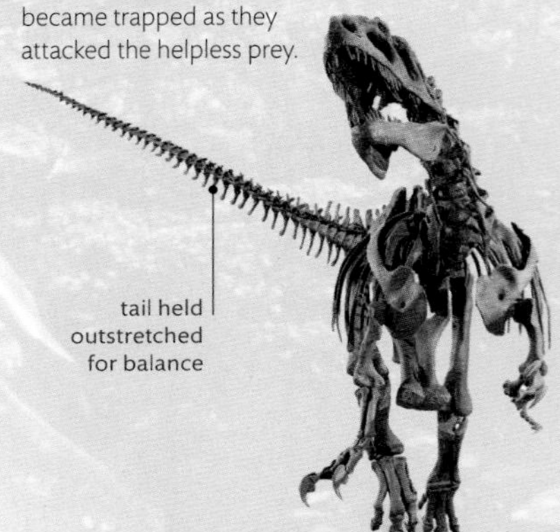

Othniel C. Marsh
One of the greatest dinosaur palaeontologist of the 19th century, Othniel C. Marsh, described *Allosaurus* and more than 500 other vertebrates.

Length: 12m (39ft)	Weight: ~2–3 tonnes (2–3 tons)	Diet: Herbivorous dinosaurs, carrion

massive skull lightened by large fenestrae (openings)
large attachment points for powerful jaw muscles
5–10cm- (2½- 5in-) long teeth
blade-like teeth for slicing
powerful upper jaw had an axe-like action

Fossil skull

Group: CARNOSAURIA	Subgroup: Allosauroidea	Time: 161–159mya

Sinraptor

SINE-RAP-TOR

Sinraptor ("Chinese thief") was part of the lineage that led to *Allosaurus*. Although it was more primitive, it had more specializations for its role as an apex predator than earlier relatives such as *Megalosaurus*. Its teeth were highly adapted for a carnivorous lifestyle: durable teeth at the front of the mouth were adapted for a bone-crunching bite, and the other teeth would have ripped through flesh to cause deep wounds. *Sinraptor* probably used these adaptations to take down giant sauropods, such as *Mamenchisaurus* (see p.77), that it lived alongside.

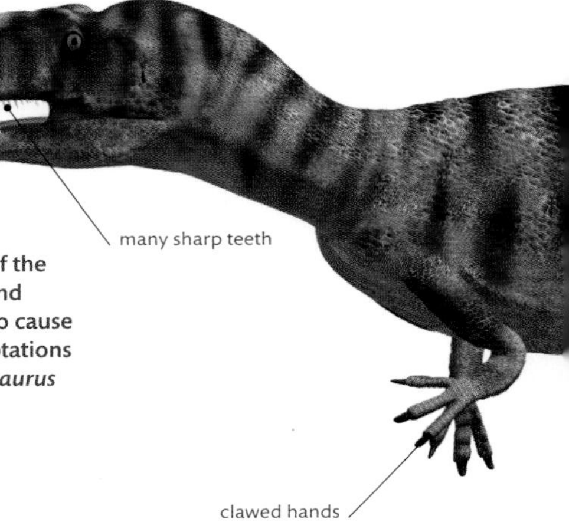

DESCRIBED BY Currie and Zhao 1993
HABITAT Marshland

Length: 7.2m (23¼ft)	Weight: ~1 tonne (1 ton)	Diet: Meat

Group: TETANURAE	Subgroup: Coelurosauria	Time: 153–149mya

Ornitholestes

OR-NITH-OH-LESS-TEES

Ornitholestes ("bird robber") was a slim, lightly-built dinosaur, with a small head, many conical teeth, and an S-shaped neck. A long, tapering tail contributed to greater agility when running. Its name arose because its grasping hands, light build, and long hind limbs made it an efficient hunter of small Jurassic animals. Its arms were short and strong; its hands had three long, clawed fingers. An additional finger was very small. *Ornitholestes* had many bird-like features, and its wrist structure allowed it to tuck its hands up close to its body in the same manner that a bird folds its wings.

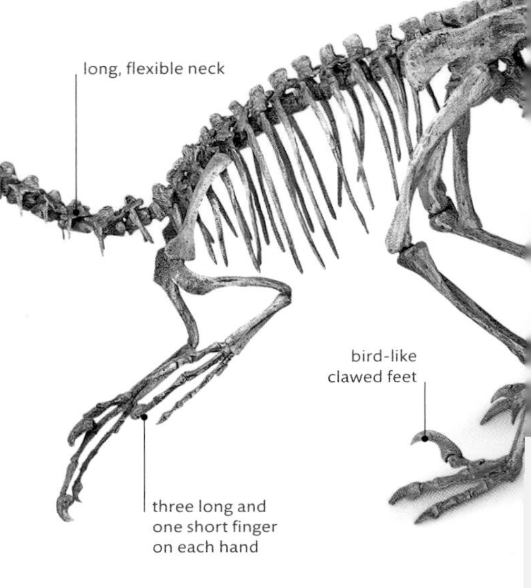

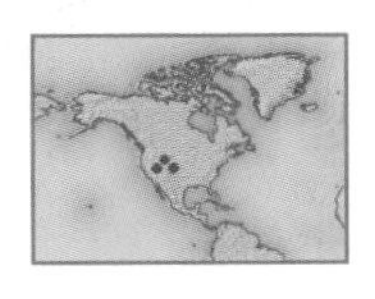

DESCRIBED BY Osborn 1903
HABITAT Forests

Length: 2m (6½ft)	Weight: ~12kg (26lb)	Diet: Meat, perhaps including birds, carrion

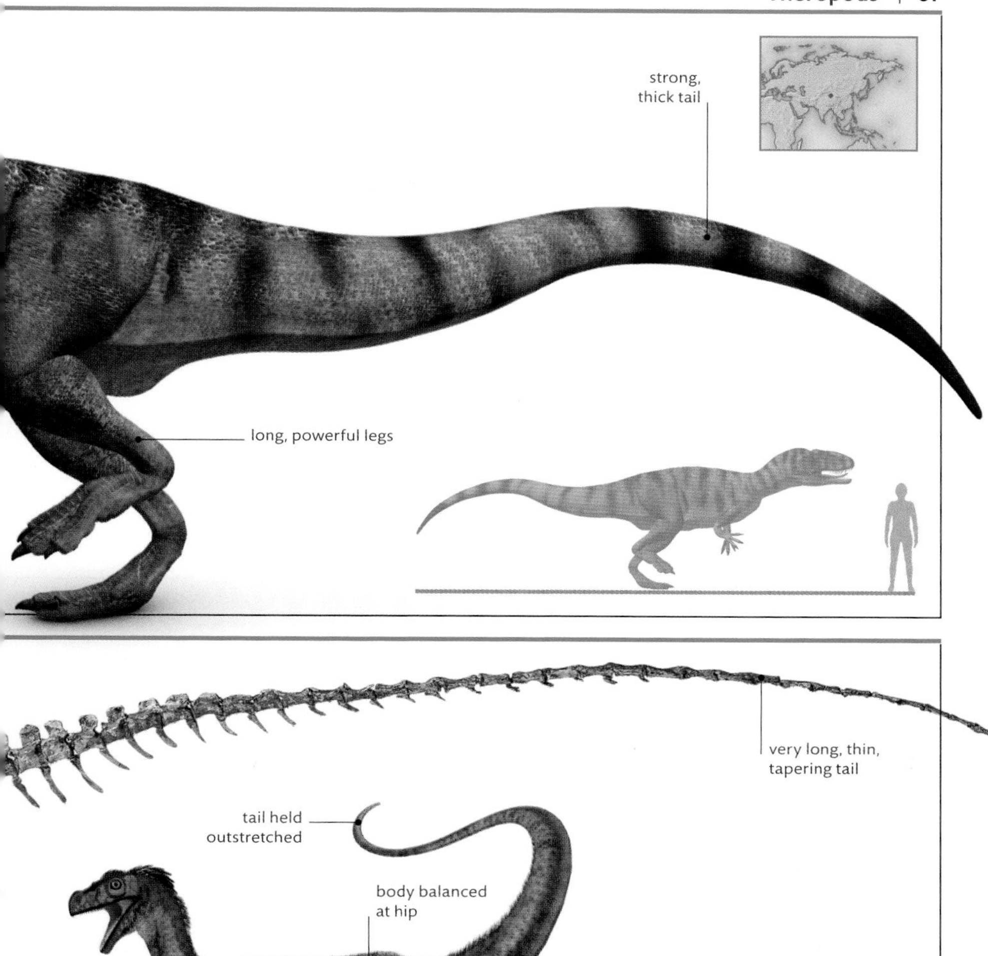

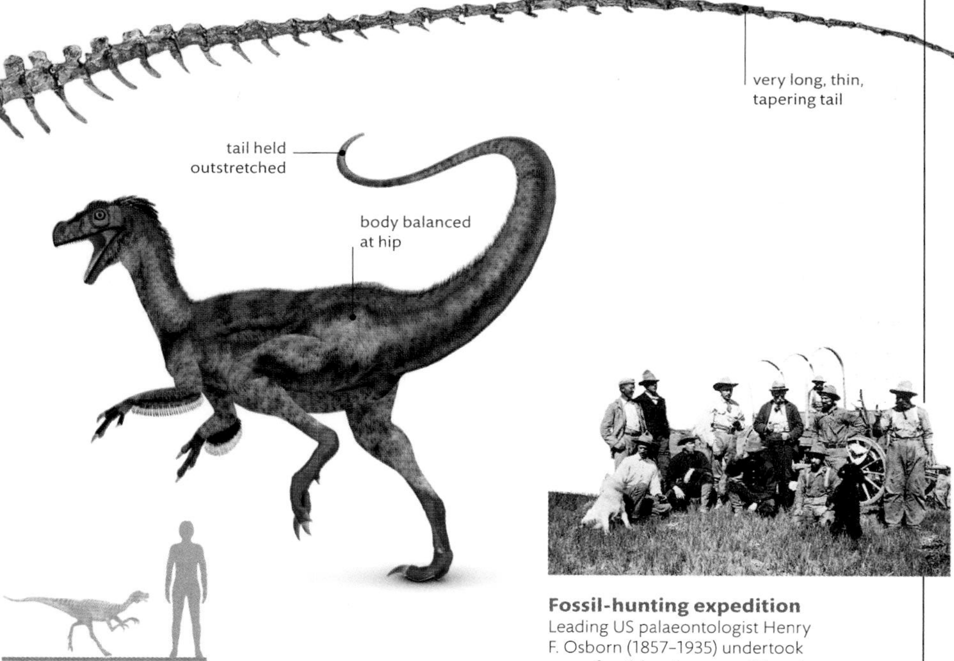

Fossil-hunting expedition
Leading US palaeontologist Henry F. Osborn (1857–1935) undertook many fossil-hunting expeditions in Mongolia and the USA in the 1900s.

Group: COELUROSAURIA	Subgroup: Compsognathidae	Time: 150mya

Compsognathus

COMP-SOG-NAY-THUS

This theropod, whose name means "pretty jaw", was about the size of a modern turkey. Its skeletal structure shows that it must have been a fast runner – its bones were hollow, its shins were much longer than its thighs, and its long tail was held outstretched for balance. Its arms were short, probably with three fingers. Its feet were very bird-like, with three clawed toes facing forwards. *Compsognathus* was anatomically very similar to *Archaeopteryx* (pp.70–71), with the exception of wings, and lived in similar areas. The presence of primitive feathers on a closely related genus – *Sinosauropteryx* – suggests that *Compsognathus* could also have had a downy body covering, although there is no direct evidence for this.

DESCRIBED BY Wagner 1861
HABITAT Warm, moist areas and scrub

ADULT REMAINS

Only two skeletons of *Compsognathus* have been found – an adult and a juvenile specimen. Both were found with a skeleton inside their stomach. At first it was thought that the remains were of a foetus, but it turned out to be a lizard (*Bavarisaurus*).

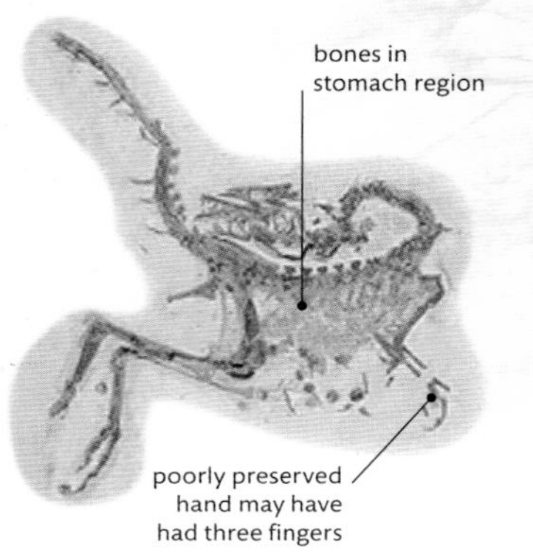

Length: 1m (3ft)	Weight: ~3kg (6½lb)	Diet: Small lizards, mammals

lightweight body
balanced at hips
tail held stiffly
out for balance
slim, short
thighs

Group: PARAVES	Subgroup: Avialae	Time: 150mya

Archaeopteryx

ARE-KAY-OP-TER-IKS

Archaeopteryx ("ancient wing") was about the size of a modern pigeon, with a small head, large eyes, and pointed teeth in its jaws. Its lower leg bones were long and slim, indicating that it could move well on land. Unlike modern birds, *Archaeopteryx* had a flat breastbone, a long, bony tail, and three grasping claws on each wing. However, its feathers, wings, and furcula (wishbone) are characteristics shared by modern birds. Recent analysis shows that *Archaeopteryx* could probably fly, although it may not have been a strong flier. It was almost certainly endothermic.

DESCRIBED BY von Meyer 1861
HABITAT Lakeshores or open forests

Length: 30cm (12in)	Weight: ~300–500g (11–18oz)	Diet: Insects

Fossil remains embedded in rock
impressions of feathers preserved in limestone
feathers attached to sides of bony tail
long, asymmetric wing feathers typical of birds capable of true flight

SAUROPODOMORPHS

THROUGHOUT THE TRIASSIC and into the Jurassic, sauropodomorphs were transitioning to a quadrupedal stance and had adopted a strictly herbivorous diet. These two features enabled them to reach much larger body sizes after the Triassic-Jurassic extinction. Some of these early Jurassic species were enormous, such as *Ledumahadi*, which at 12 tonnes (13.2 tons) would have been the largest land animal ever at that time.

In the early Jurassic, the first true sauropods appeared, distinguished by their pillar-like legs and long necks with tiny heads. Sauropods would become incredibly successful in the Jurassic, radiating around the globe and reaching truly astonishing sizes. Three main groups diverged, and each developed a unique body plan, which allowed them to coexist throughout the Jurassic. The diplodocids had elongated necks and tails, and were able to grow to incredibly large sizes. Diplodocids include some of the best known sauropods, such as *Apatosaurus* and the recently resurrected *Brontosaurus*. Their close cousins, the dicraeosaurs, remained relatively small and had shorter necks adorned with two rows of tall spines. The third group, the macronarians, had longer forelimbs that gave them a more upright stance, as exemplified by *Giraffatitan* (see pp.80–81). Macronarians, named for their large nostrils, varied in size, with some sub-groups, such as camarasaurids, remaining small, and others, such as the brachiosaurids and later titanosaurians, becoming much larger.

Group: SAUROPODOMORPHA	Subgroup: Massospondylidae	Time: 201–182mya

Massospondylus

MASS-OH-SPON-DI-LUSS

Massospondylus ("massive vertebrae") was named for the first remains found - a few large vertebrae. Many fossil finds have now been made, and it appears to have been the most common sauropodomorph in what is now southern Africa. *Massospondylus* had a tiny head on a long, flexible neck. It was previously thought to have spent most of its time on all fours, but recent analysis suggests that *Massospondylus* moved around on its hind legs. It had massive, five-fingered hands that could be used for holding food. Each thumb had a large, curved claw. The front teeth were rounded, and the back ones had flat sides, indicating a tough plant diet. Gizzard stones have been found with some skeletons.

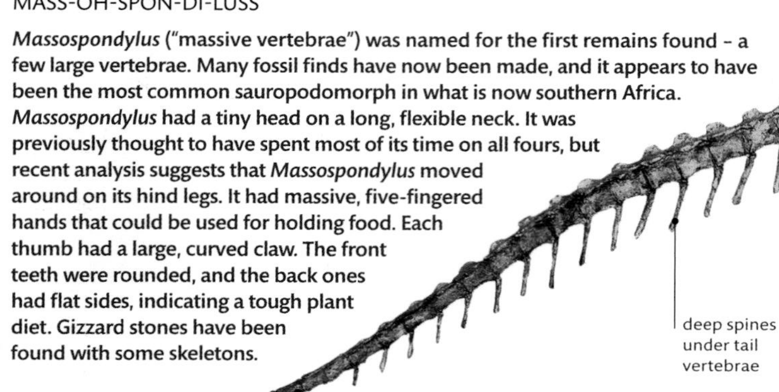

DESCRIBED BY Owen 1854
HABITAT Scrubland and desert plains

Length: 4m (13ft)	Weight: ~150kg (330lb)	Diet: Plants, possibly small animals

Group: SAURISCHIA	Subgroup: Sauropodomorpha	Time: 201–190mya

Anchisaurus

AN-CHEE-SAW-RUS

***Anchisaurus* was an early sauropodomorph. It had a small head with a narrow snout, a long, flexible neck, and a slim body and tail. Although its arms were a third shorter than its legs, it spent most of its time on all fours. The thumbs had a large claw that may have been used for defence.**

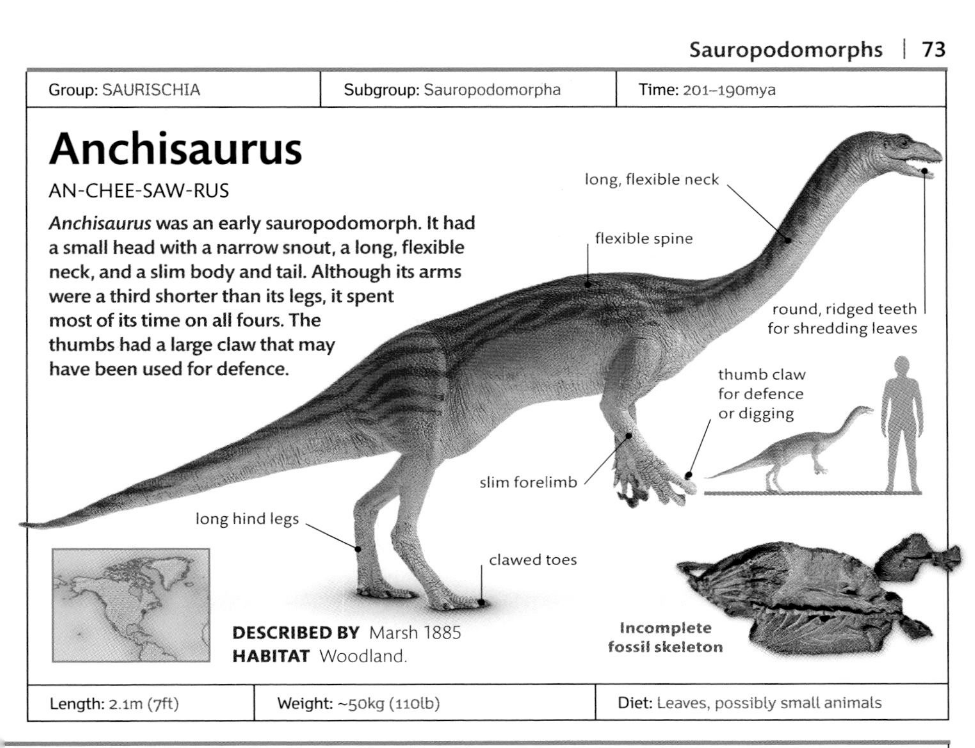

Incomplete fossil skeleton

DESCRIBED BY Marsh 1885
HABITAT Woodland.

Length: 2.1m (7ft)	Weight: ~50kg (110lb)	Diet: Leaves, possibly small animals

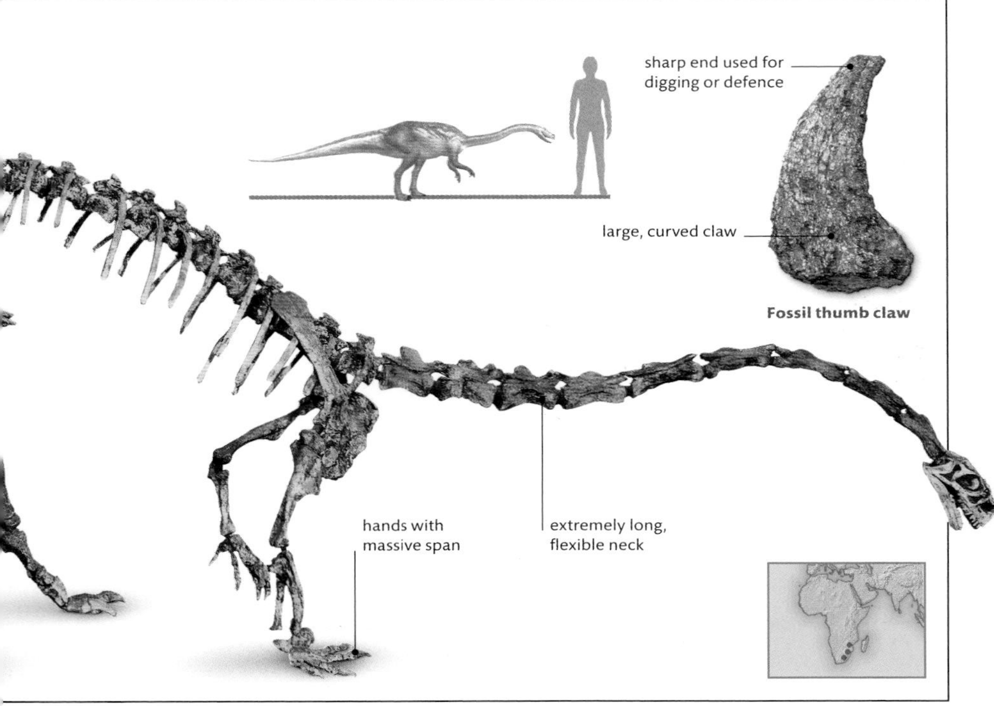

Fossil thumb claw

Group: SAUROPODOMORPHA	Subgroup: Massospondylidae	Time: 201–190mya

Lufengosaurus

LOO-FUNG-OH-SAW-RUS

Lufengosaurus ("lizard from Lufeng") was a heavy, stout-limbed sauropodomorph. Its small head held many widely spaced teeth that were shaped like leaves, a typical feature of sauropodomorph dinosaurs. Its lower jaw was hinged below the level of the upper teeth. This gave the jaw muscles greater leverage for feeding on tough plant material. Its broad feet had four long toes, and its large hands had long, clawed fingers; the thumbs bore a massive claw. *Lufengosaurus* would have moved around on its hind legs, sometimes rearing up and stretching out its long neck to feed on cycad or conifer trees.

DESCRIBED BY Young 1940
HABITAT Desert plains

relatively large, deep head

paired pubic bones projected back

bulky, heavy body

hands with large span

Length: 6m (20ft)	Weight: ~1–3 tonnes (1–3 tons)	Diet: Cycad and conifer leaves

Group: SAUROPODOMORPHA	Subgroup: Sauropoda	Time: 199–182mya

Barapasaurus

BA-RAP-A-SAW-RUS

Barapasaurus ("big-legged lizard") is one of the earliest known sauropods. There is fossil evidence for all parts of the skeleton, with the exception of the skull and feet, making it potentially the best-known Early Jurassic sauropod. However, to date it still has not been fully described, and its provisional classification may change in the light of future findings. *Barapasaurus* had slim limbs, spoon-shaped, saw-edged teeth (found in isolation), and unusual hollows in the vertebrae of the back. Palaeontologists believe this dinosaur had a short, deep head, like skulls from other primitive sauropods.

DESCRIBED BY Jain et al. 1975
HABITAT Plains

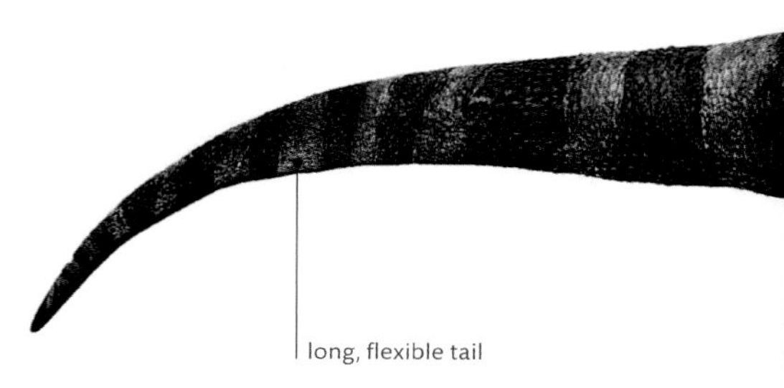

Length: 18m (59ft)	Weight: ~20 tonnes (19⅝ tons)	Diet: Plants

Group: SAUROPODOMORPHA	Subgroup: Sauropoda	Time: 159mya

Shunosaurus

SHOON-OH-SAW-RUS

Nearly complete skeletons of *Shunosaurus* ("Shu lizard") have been discovered, and it was only the second sauropod to be known in its entirety. The skull is long and low with small teeth. A surprising feature is the small bony club at the end of its tail, formed by enlarged vertebrae (not seen in the pictured specimen).

DESCRIBED BY
Dong et al. 1983
HABITAT Plains

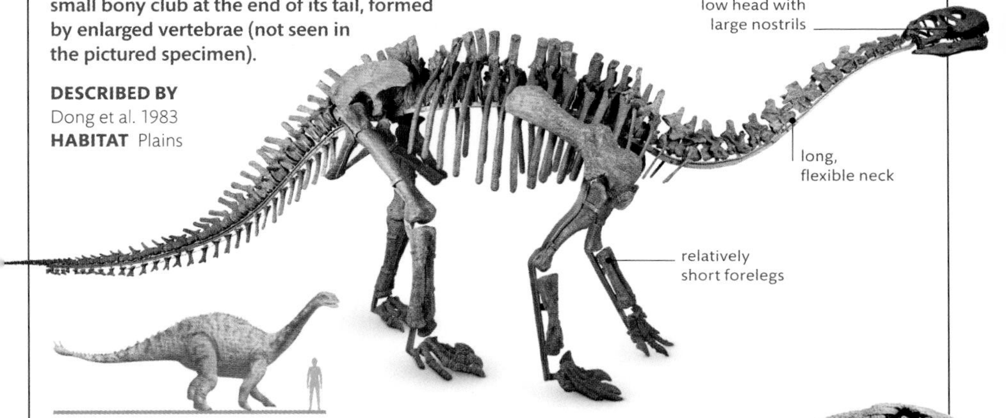

low head with large nostrils

long, flexible neck

relatively short forelegs

Length: 10m (33ft)	Weight: ~10.1 tonnes (10 tons)	Diet: Plants

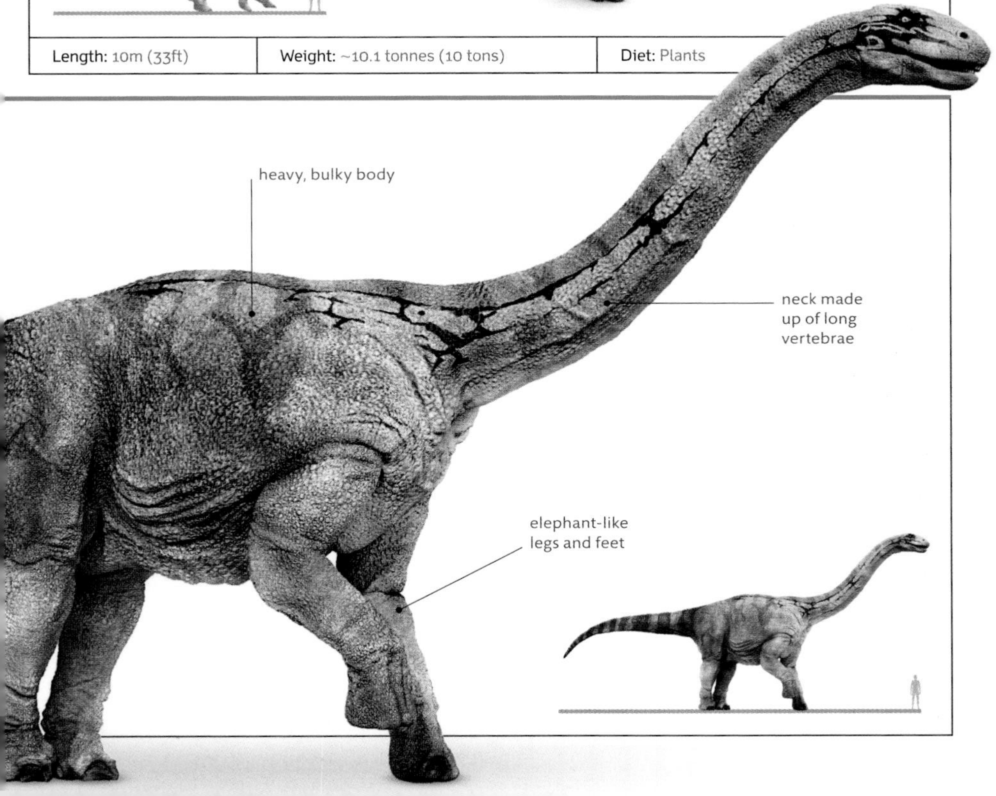

heavy, bulky body

neck made up of long vertebrae

elephant-like legs and feet

Group: SAUROPODA	Subgroup: Diplodocidae	Time: 152–150mya

Barosaurus

BA-ROH-SAW-RUS

Barosaurus ("heavy lizard") had all the typical features of its family – a long neck and tail, a bulky body, tiny head, and relatively short legs for its size. Its limbs are indistinguishable from those of *Diplodocus* (see p.78), but its neck was much longer. Its 15 cervical vertebrae were greatly elongated – a third longer than in *Diplodocus*. *Barosaurus* probably roamed in herds and relied on its huge size for defence against the large predators of the time.

DESCRIBED BY Marsh 1890
HABITAT Floodplains

Dinosaur hunting
The camp at Carnegie Quarry, USA, where three *Barosaurus* skeletons were found in 1922.

Length: 23–27m (75–89ft)	Weight: ~19⅝ tonnes (20 tons)	Diet: Plants

Group: SAUROPODOMORPHA	Subgroup: Sauropoda	Time: 168–166mya

Cetiosaurus

SEE-TEE-OH-SAW-RUS

This dinosaur, whose name means "whale lizard", was discovered early in the 18th century. Its huge bones were thought to belong to some type of immense whale – hence its name. It was a large, heavy sauropod with a shorter neck and tail than other sauropods. It was also different in that its vertebrae had fewer hollow spaces to lighten them than other sauropods of its size. Its head was blunt, and contained spoon-shaped teeth. *Cetiosaurus* is thought to have roamed in large herds, at an estimated walking speed of 10mph (15kph).

DESCRIBED BY
Owen 1841
HABITAT Plains

solid centrum of vertebra

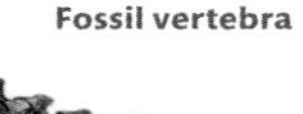

Fossil vertebra

tail held outstretched in life

1.5m- (5ft-) long shoulder blade (mounted too vertically in this reconstruction)

Length: 18m (59ft)	Weight: ~9 tonnes (8⅞ tons)	Diet: Plants

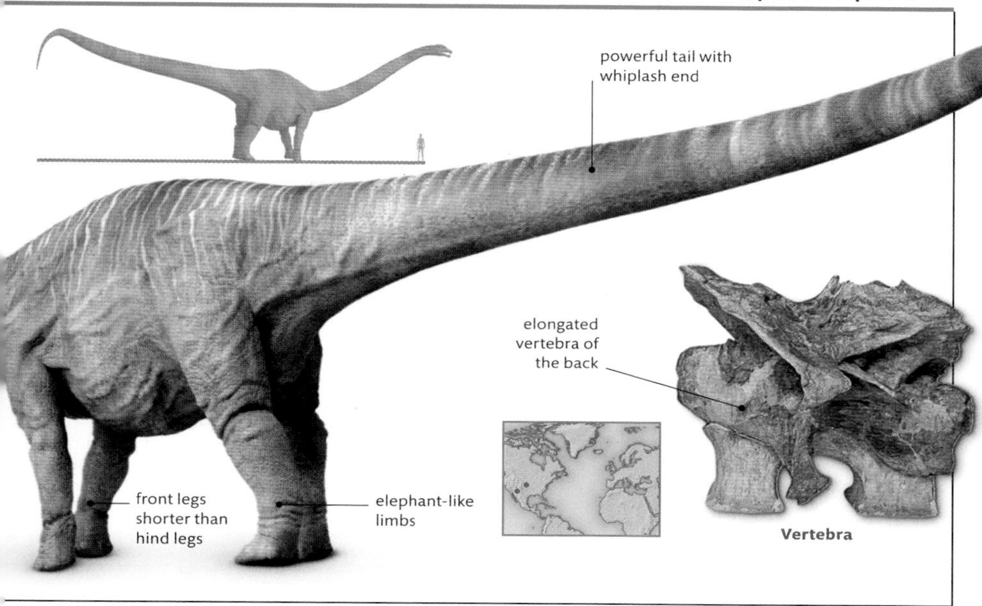

Group: SAUROPODOMORPHA	Subgroup: Sauropoda	Time: 152–114mya

Mamenchisaurus

MAH-MEN-CHI-SAW-RUS

This dinosaur, whose name means "Mamenxi lizard", had one of the longest necks of any dinosaur. It made up more than half of the animal's total length, and contained 19 vertebrae – more than any other dinosaur. The neck vertebrae were over twice the length of those in its back, and were overlapped by long, thin, bony struts. It is thought that this restricted movement to the joint between the head and the uppermost neck bone, with the dinosaur holding its neck not much higher than shoulder height. It could then swing it widely from the shoulders to browse on low vegetation.

Fossil remains
Remains of *Mamenchisaurus* have been excavated from sites in China.

back sloping from shoulders

enormously elongated neck

DESCRIBED BY Young 1954
HABITAT Deltas and forested areas

Length: 35m (114ft)	Weight: ~60 tonnes (66 tons)	Diet: Leaves and shoots

Group: SAUROPODA	Subgroup: Diplodocidae	Time: 154–152mya

Diplodocus

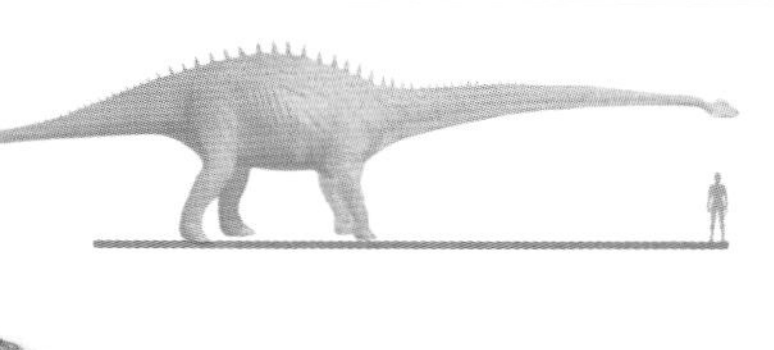

DIPLOH-DOCK-US

Diplodocus was one of the longest dinosaurs. It was lightly built, with slim limbs and a tapered tail with a whiplash end. Its name meaning "double beam", comes from the chevron bones under its tail vertebrae. There are three species of *Diplodocus: D. longus, D. carnegii*, and *D. hallorum*.

peg-like teeth at front of jaws

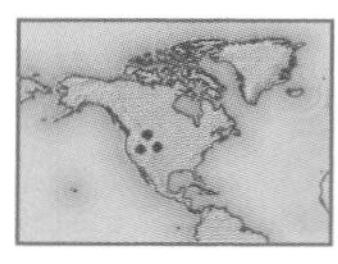

DESCRIBED BY Marsh 1878
HABITAT Plains

Length: 32m (104ft)	Weight: ~20 tonnes (22 tons)	Diet: Leaves

Group: SAUROPODA	Subgroup: Macronaria	Time: 156–149mya

Camarasaurus

CAM-A-RAH-SAW-RUS

Camarasaurus ("chambered lizard") was the most common sauropod in North America. It appears to have roamed in large herds. This dinosaur had a relatively large, box-shaped head and short neck and tail. The vertebrae probably contained air spaces to lighten the backbone. The front legs of *Camarasaurus* were relatively long. Its forefeet had one single claw, and the hindfeet three.

DESCRIBED BY Cope 1877
HABITAT Plains

Length: 23m (75ft)	Weight: ~47 tonnes (51 tons)	Diet: Tough vegetation

Group: SAUROPODA	Subgroup: Diplodocidae	Time: 152–151mya

Apatosaurus

A-PAT-OH-SAW-RUS

Apatosaurus was shorter, yet heavier and bulkier than its close relatives. It had a tiny head at the end of a long neck made up of 15 vertebrae. Its back vertebrae were hollow, and its long tail had a whip-like end. Its thick hind legs were longer than the front ones, perhaps to allow the dinosaur to rear up when feeding. Whether *Apatosaurus* could actually do this is disputed by some palaeontologists.

DESCRIBED BY Marsh 1877
HABITAT Wooded plains

neck with limited mobility

tail comprising over half the total length

relatively short front legs

Length: 21m (70ft)	Weight: ~37 tonnes (42 tons)	Diet: Leaves

Fossil skull

Group: SAUROPODA	Subgroup: Macronaria	Time: 157–145mya

Giraffatitan

JI-RAF-A-TIE-TAN

Giraffatitan is known from all parts of the skeleton except for the important neural arches of the vertebrae at the base of the neck. It was one of the tallest and largest sauropods, and the one in which the lengthening of the forelimbs relative to the hindlimbs reached its extreme. Together with its long neck, ending in a small head, *Giraffatitan's* giraffe-like stance gave it a great height: up to 16m (50ft). The tail was relatively short and thick. Like other members of its family, *Giraffatitan* had chisel-like teeth - 26 on each jaw - towards the front of the mouth. The nostrils were at the front of the snout and soft tissue ran back to the nostril openings in the skull. The legs were pillar-like, and the feet all had five toes with fleshy pads behind. The first toe of each front foot bore a claw, as did the first three toes of the hind feet. Like other sauropods, it probably travelled in herds. *Giraffatitan* was adapted for feeding off vegetation from the tops of trees. The length of its neck means that it would have had to have an extremely large, powerful heart, and very high blood pressure to pump blood up to its head. *Giraffatitan* is thought to have laid its eggs while walking, leaving the young to fend for themselves.

DESCRIBED BY Janensch 1914
HABITAT Plains

MASSIVE BONES

The femur (thigh bone) of the closely-related *Brachiosaurus* was over 1.8m (6ft) long and massively thick to support the dinosaur's weight. One of the first bones discovered by Elmer S. Riggs, in 1900, was the humerus (upper arm bone). It was over 2.1m (7ft) long, which was so much greater than that of any known humerus, that Riggs thought it was a crushed femur of an *Apatosaurus* (see p.79).

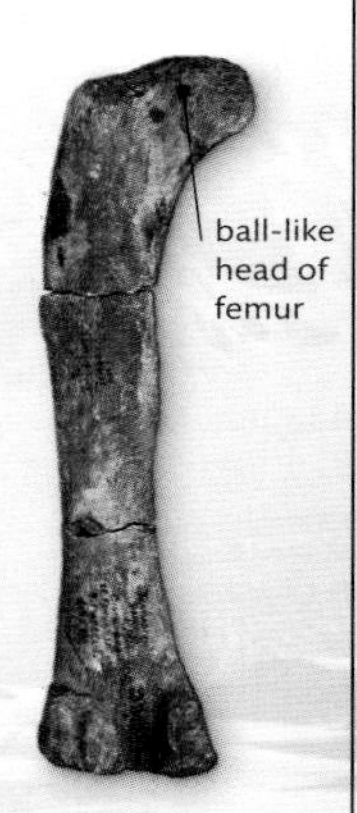

body sloping downwards from shoulder to hip

short tail

relatively slim, pillar-like legs

Length: 26m (85ft)	Weight: ~50 tonnes (49½ tons)	Diet: Plants

chisel-like teeth
at front of mouth
long neck made
up of 1m- (3ft-)
long vertebrae
air sacs in
the vertebrae
lightened the neck
forelimbs longer
than hindlimbs

ORNITHISCHIANS

THE ORIGINS of the herbivorous ornithischians are still a mystery. There are no definitive records prior to the Jurassic, even though they should have branched off from other dinosaurs during the Late Triassic.

Ornithischians had a toothless, bony beak at the front of the mouth, and cheeks that would have held food as it was ground by specialized teeth. Two lineages of ornithischians are known from the early part of the Jurassic: the armoured thyreophorans and the agile, bipedal cerapodans. Early thyreophorans, such as *Scutellosaurus*, were bipedal, but the group would soon become quadrupedal and split into the plate-backed stegosaurs and heavily armoured ankylosaurs. Throughout the Jurassic these two groups shared the role of large herbivores with the sauropods, although they relied on their armour and weaponized tails for protection rather than reaching the enormous sizes of sauropods.

The exact relationships of early cerapodans are still being untangled, but by the Middle Jurassic, representatives of both lineages – marginocephalians and ornithopods – were known. The marginocephalians – named for the extended margin protruding from the back of their skull, would give rise to frilled ceratopsians and dome-headed pachycephalosaurs. For most of the Jurassic, ornithopods remained small, quick, bipedal runners, but some started to get larger later – foreshadowing the rise of hadrosaurs and iguanodontians in the Cretaceous.

Group: DINOSAURIA	Subgroup: Ornithischia	Time: 201–190mya

Lesothosaurus

LE-SOO-TOE-SAW-RUS

Lesothosaurus ("Lesotho lizard") was an early bipedal herbivorous dinosaur. It was small and built for speed, with a lightly-built body, long, slim legs, and a thin, flexible tail. It had pointed front teeth, serrated, arrow-shaped cheek teeth, and its jaws moved only up and down, not from side to side. *Lesothosaurus* had a limited ability to chew its food, and most of the food processing was done with its beak. It may have been an omnivore.

DESCRIBED BY Galton 1978
HABITAT Desert plains

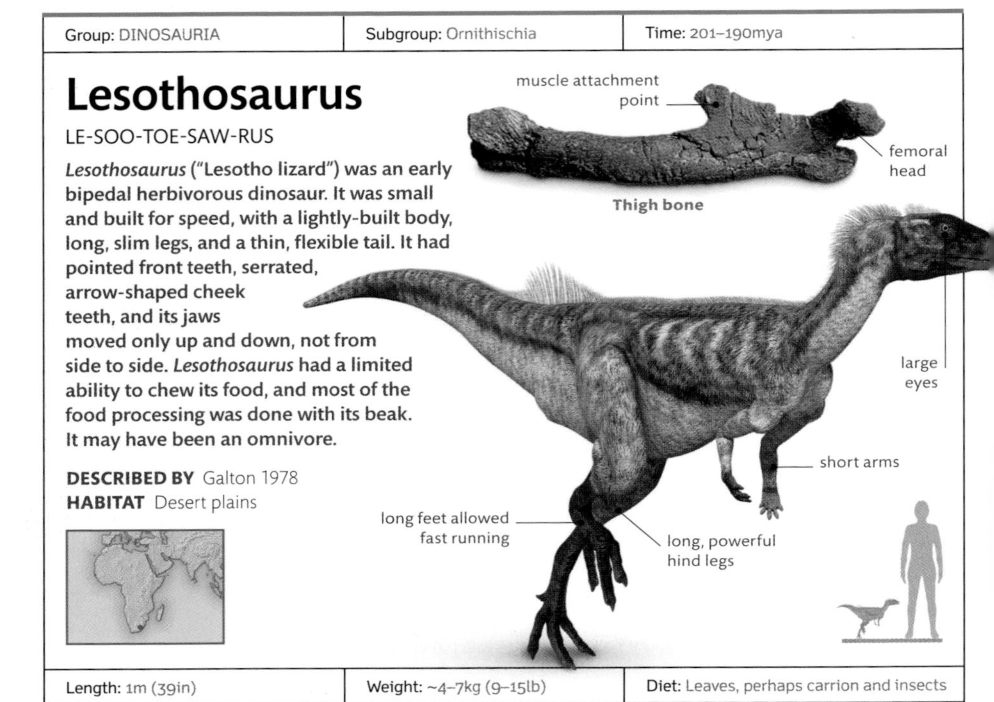

Length: 1m (39in)	Weight: ~4–7kg (9–15lb)	Diet: Leaves, perhaps carrion and insects

Group: ORNITHISCHIA	Subgroup: Thyreophora	Time: 199–190mya

Scutellosaurus

SCOO-TELL-OH SAW-RUS

This "little shield lizard" had a long body and slim limbs. Its arms were also relatively long and slender. Over 300 low, bony studs covered its back, flanks, and the base of its tail to form a defensive armour. Recent analysis suggests that *Scutellosaurus* spent most of its time on its hind legs, holding its tail stiffly out behind it to counterbalance the weight of the bony armour.

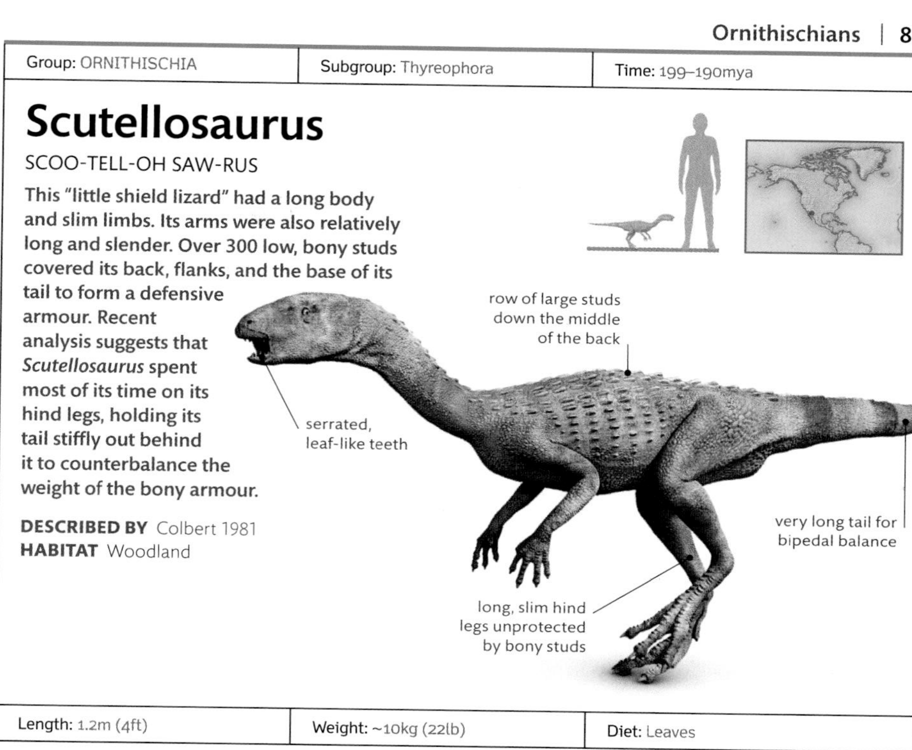

DESCRIBED BY Colbert 1981
HABITAT Woodland

Length: 1.2m (4ft)	Weight: ~10kg (22lb)	Diet: Leaves

Group: THYREOPHORA	Subgroup: Ankylosauria	Time: 156–149mya

Gargoyleosaurus

GAR-GOYL-EE-OH-SAW-RUS

***Gargoyleosaurus* "gargoyle lizard" was a primitive, early ankylosaur with body armour formed from irregular, bony oval plates on the upper surface of its body and tail. The hind legs were slightly longer than the front ones. *Gargoyleosaurus* had many features unusual among ankylosaurs, including a straight nasal passage and hollow armour plates. The skull shows features seen in late Cretaceous ankylosaurs, including fusion of bone armour to the surface of the skull and jaw.**

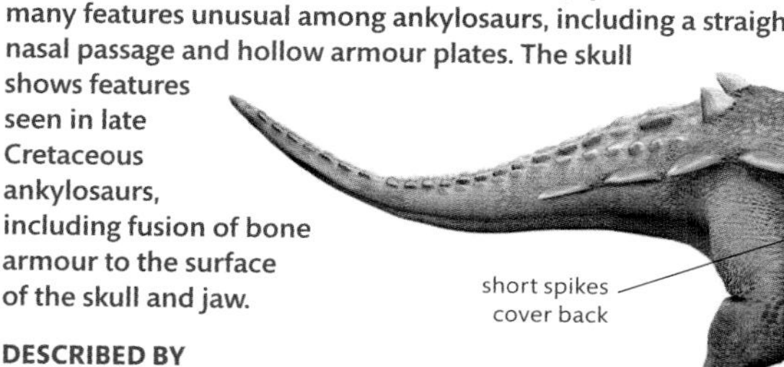

DESCRIBED BY
Carpenter et al. 1998
HABITAT Woodland

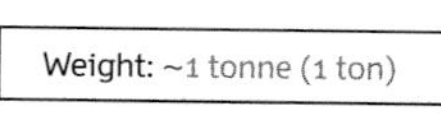

Length: 3m (9¾ft)	Weight: ~1 tonne (1 ton)	Diet: Low-lying vegetation

Group: ORNITHISCHIA	Subgroup: Thyreophora	Time: 199–182mya

Scelidosaurus

SCEL-EYE-DOH-SAW-RUS

This small, heavy, armoured dinosaur, whose name means "limb lizard", seems to be one of the earliest and most primitive thyreophorans. Recent research suggests that it was a primitive ankylosaur. It had impressive defensive armour: its back was covered with bony plates, and studded with a double row of bony spikes. Additional rows of studs ran along its sides, and there was a pair of triple-spiked bony plates behind the neck. The head was small and pointed with a horny beak, and small, leaf-shaped teeth. Although the forelimbs were much shorter than the hind limbs, *Scelidosaurus* seems to have walked on all fours.

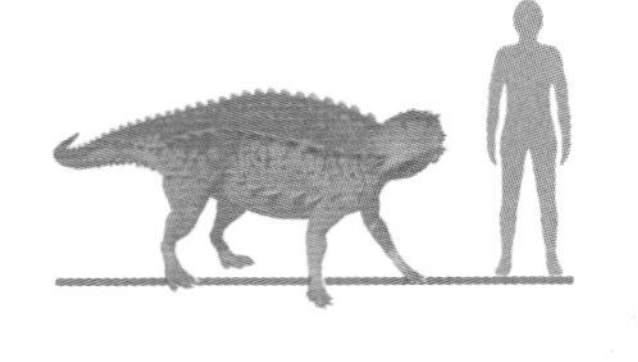

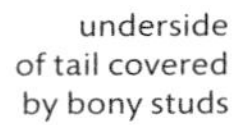

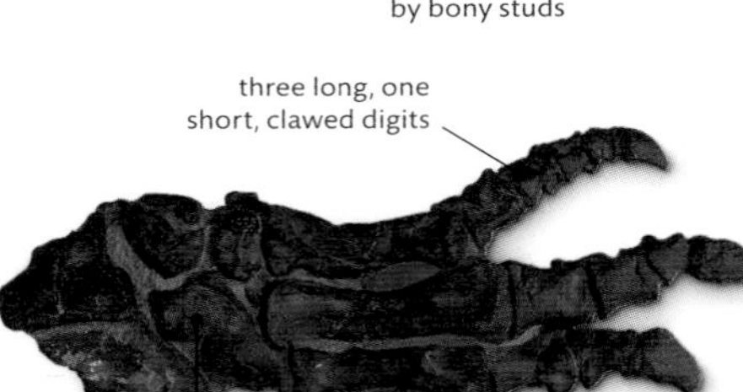

Fossil foot

DESCRIBED BY Owen 1861
HABITAT Woodland

Length: 3.5m (11ft)	Weight: ~250kg (550lb)	Diet: Plants

Group: THYREOPHORA	Subgroup: Stegosauria	Time: 156–150mya

Kentrosaurus

KEN-TRO-SAW-RUS

This East African contemporary of *Stegosaurus* was smaller than its more famous relative, but as well protected. *Kentrosaurus* ("spiked lizard") had paired rectangular plates running down over the neck, shoulders, and half of the back. Over the hips, the plates gave way to sharp spikes, which ran in pairs down to the tip of the tail. Another pair of long spikes jutted out from the hips on each side. The skull is known only from fragments, but was probably long and slender with a toothless, horny beak.

DESCRIBED BY Hennig 1915
HABITAT Forests

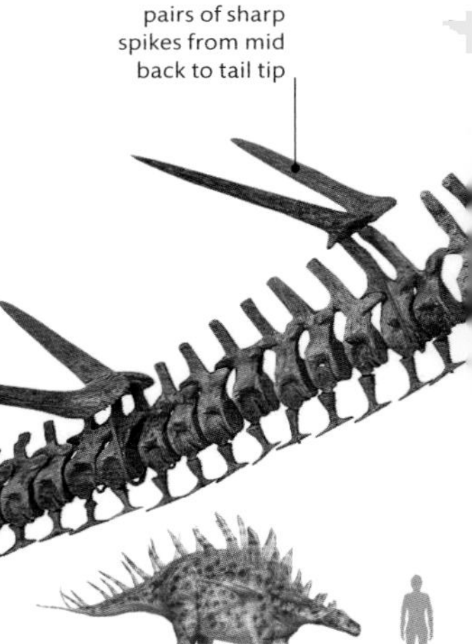

Length: 5m (16ft)	Weight: ~1.5 tonnes (1½ tons)	Diet: Low-lying vegetation

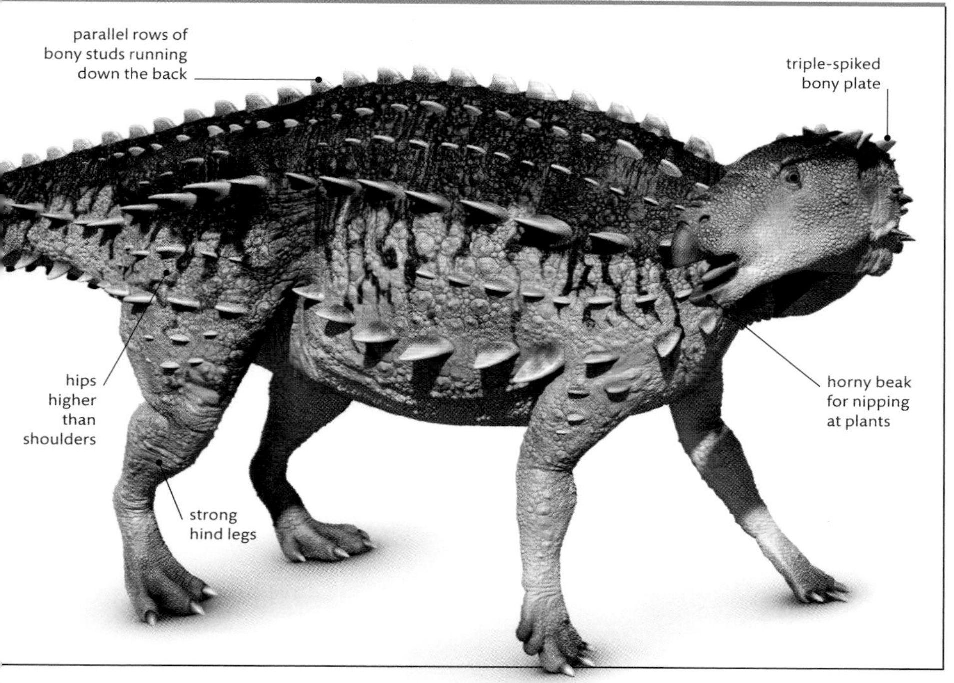
parallel rows of
bony studs running
down the back
triple-spiked
bony plate
hips
higher
than
shoulders
horny beak
for nipping
at plants
strong
hind legs

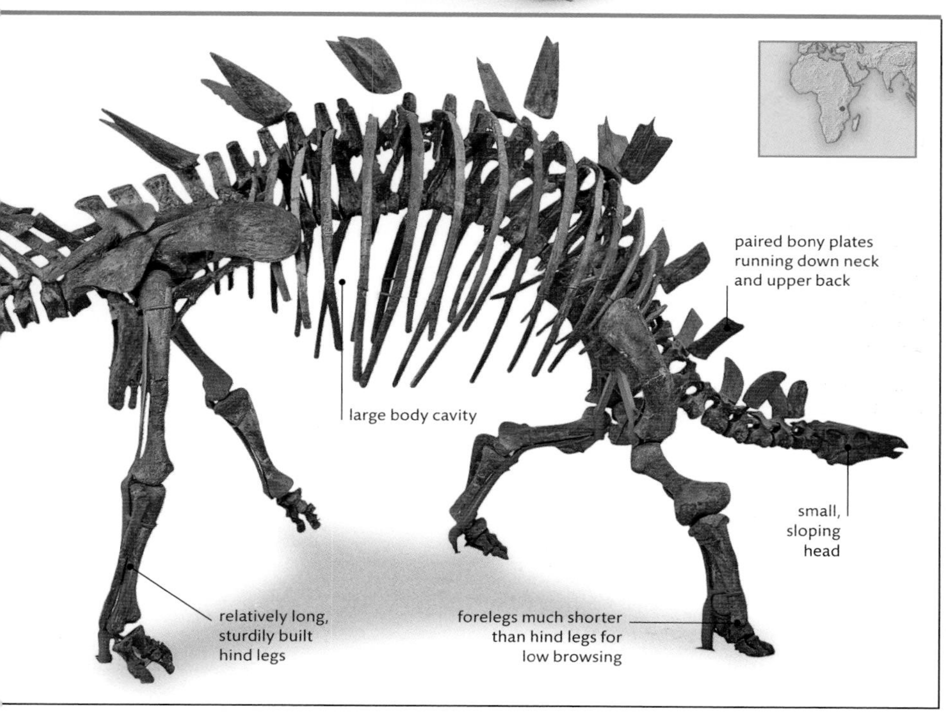
paired bony plates
running down neck
and upper back
large body cavity
small,
sloping
head
relatively long,
sturdily built
hind legs
forelegs much shorter
than hind legs for
low browsing

Group: THYREOPHORA	Subgroup: Stegosauria	Time: 156–149mya

Stegosaurus

STEG-OH-SAW-RUS

Stegosaurus ("roof lizard") was the largest plated dinosaur. It had a tiny head, toothless beak, and two rows of bony plates running from behind its neck to halfway along its tail. Some of these plates were over 60cm (24in) high. The tail also had long spikes, each up to 1m (39in) long, the number of which varied between species. The dinosaur's neck was also armoured with many tiny bony studs. The hind legs were twice as long as the forelegs, meaning that the body's highest point was at the hips. *Stegosaurus* had a toothless beak and small cheek teeth that could not chew, so exactly how *Stegosaurus* processed enough vegetation to survive remains unclear.

DESCRIBED BY Marsh 1877
HABITAT Woodland

Length: 6m (20ft)	Weight: ~2 tonnes (2 tons)	Diet: Plants

Skeletal reconstruction

PLATES AND SPIKES

Some early reconstructions of *Stegosaurus* have the plates arranged in pairs. Scientists have revised their reconstructions over time to put them in a zigzag pattern.

Fossil plate

Fossil tail spike

Group: ORNITHISCHIA	Subgroup: Heterodontosauridae	Time: 201–190mya

Heterodontosaurus

HET-ER-OH-DONT-OH-SAW-RUS

As its name ("different-toothed lizard") suggests, the most remarkable feature of this small bipedal dinosaur was its teeth. *Heterodontosaurus* had three kinds: cutting incisors at the front of the upper jaw, two pairs of large tusk-like teeth, and tall chisel-like teeth used for shredding vegetation. The tusks may have been used for defence, sexual display, or feeding.

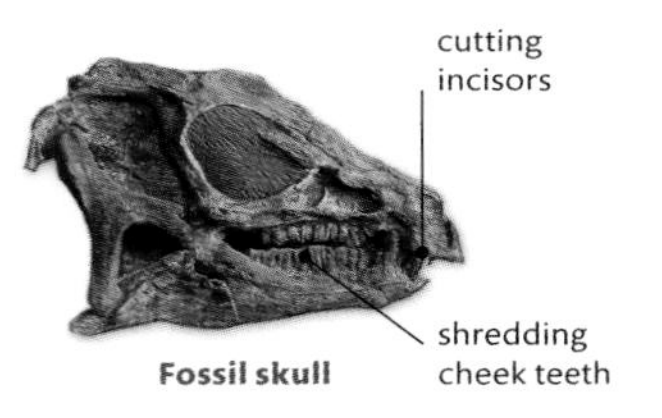

Fossil skull

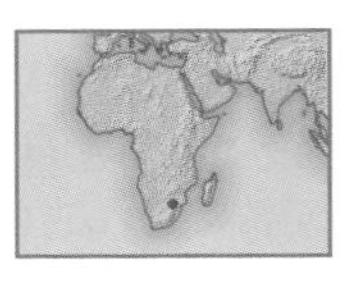

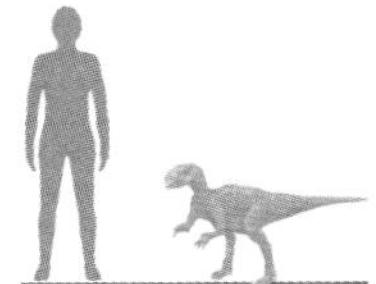

DESCRIBED BY Crompton and Charig 1962
HABITAT Scrub

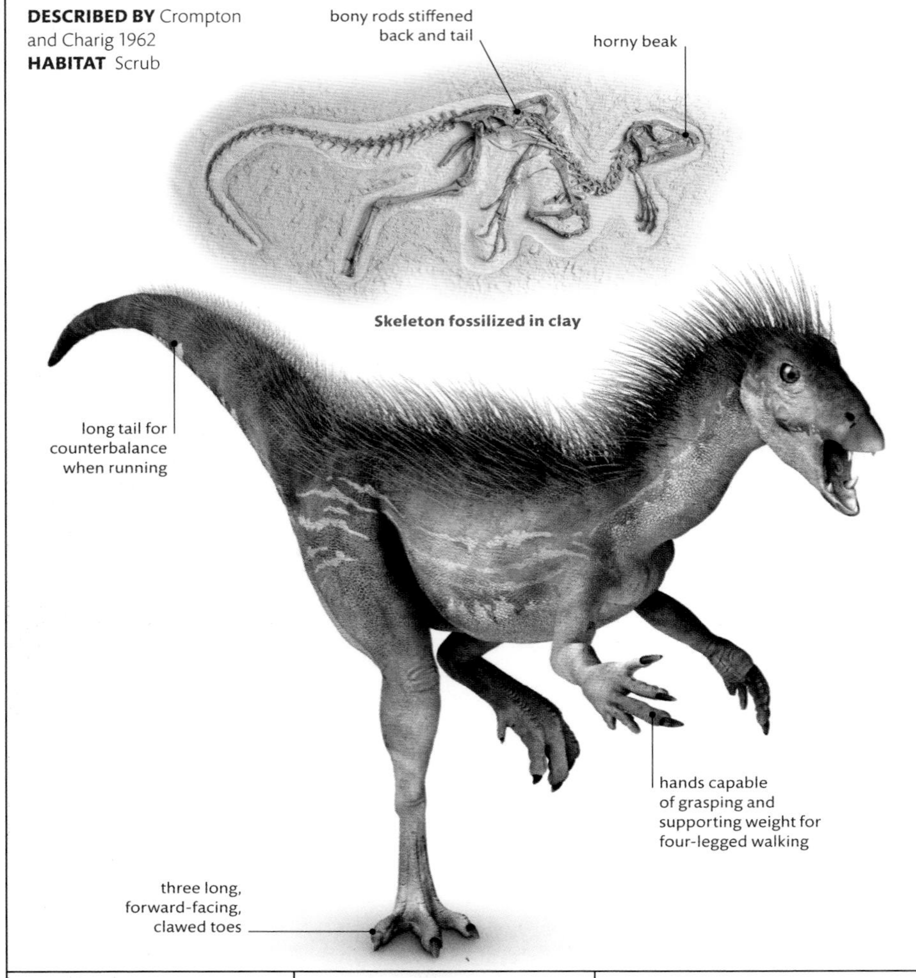

Skeleton fossilized in clay

Length: 1.2m (4ft)	Weight: ~19kg (42lb)	Diet: Leaves, tubers, possibly insects

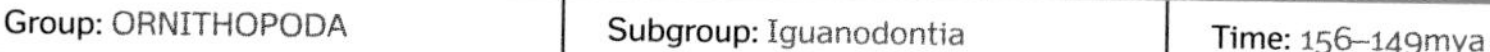

Group: ORNITHOPODA	Subgroup: Iguanodontia	Time: 156–149mya

Dryosaurus

DRY-OH-SAW-RUS

Dryosaurus ("oak lizard") was a lightly built bipedal herbivore, with powerful, slender legs that were much longer than the arms. It was a swift runner. The tail was stiffened by bony tendons for better balance. The horny beak at the front of the lower jaw met with a tough, toothless pad on the lower jaw – a perfect arrangement for cropping tough vegetation.

DESCRIBED BY Marsh 1894
HABITAT Woodland

Thigh bone

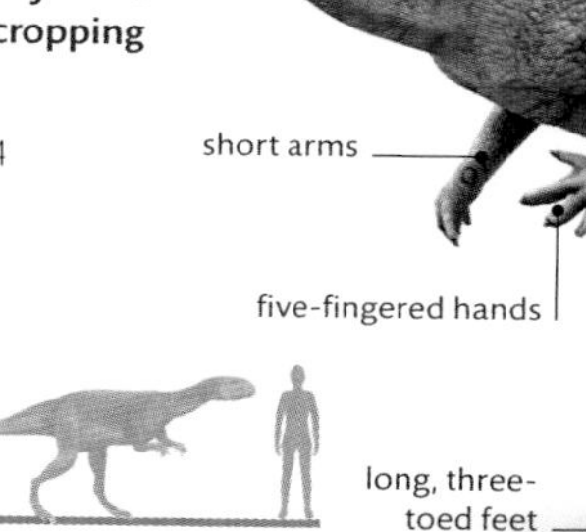

Length: 3–4m (10–13ft)	Weight: ~80kg (175lb)	Diet: Leaves and shoots

Group: ORNITHOPODA	Subgroup: Iguanodontia	Time: 156–149mya

Camptosaurus

KAMP-TOE-SAW-RUS

Camptosaurus ("bent lizard") was a bulky plant-eater that browsed on plants and shrubs close to the ground. Its head was long and low, with a sharp, horny beak at the tip of the broad snout. The pubis was shifted back to allow more space for the intestines. Its arms were shorter than its legs, with a large wrist and hoof-like claws on its fingers. *Camptosaurus* was bipedal, but often walked on its forelimbs when feeding.

sigmoid curve in neck indicating head was held low

four-toed hind feet

tail stiffened by bony ligaments

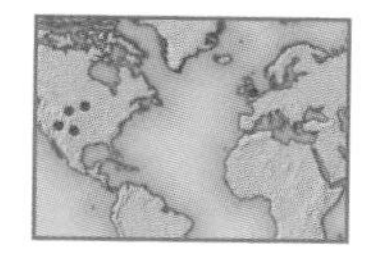

DESCRIBED BY Marsh 1885
HABITAT Open woodland

Length: 5–7m (16–23ft)	Weight: ~750kg (¾ ton)	Diet: Low-growing herbs and shrubs

OTHER DIAPSIDS

ALTHOUGH DINOSAURS ruled the Jurassic lands, their diapsid relatives took over the seas and skies.

Pseudosuchians had been displaced from their position atop terrestrial ecosystems, and all but two lineages went extinct in the Triassic-Jurassic extinction. One of these, the sphenosuchians, survived as agile terrestrial forms until the end of the Jurassic, while the other, the crocodyliformes, are still alive today. Crocodyliformes were much more diverse in the Jurassic and Cretaceous than they are today, and they radiated into a diverse array of niches, including marine predators and even terrestrial herbivores.

The marine thalattosuchians shared the seas with ichthyosaurs and plesiosaurs. Ichthyosaurs were successful at the beginning of the Jurassic and they became highly specialized for life in the ocean, developing streamlined bodies, large eyes, and even layers of blubber for insulation similar to modern whales and dolphins. Some of the carnivorous plesiosaurs became enormous in the middle of the Jurassic and they displaced ichthyosaurs as the top predators of the seas.

The flying pterosaurs became exceptionally diverse in the Jurassic and spread worldwide. Most Jurassic pterosaurs remained small and retained a mouthful of conical teeth for catching insects or fish. Towards the end of the Jurassic, the pterodactyloids arose. They lacked the long tail of earlier pterosaurs, were able to reach much larger sizes, and many lost their teeth later in the Cretaceous.

Group: ICHTHYOPTERYGIA	Subgroup: Ichthyosauria	Time: 201–174mya

Temnodontosaurus

TEM-NOH-DON-TOH-SAW-RUS

This large ichthyosaur looked a bit like a modern dolphin. Its body was long, smooth, and streamlined, and it had a long, narrow snout, with many large teeth set into a dental groove. Its tail was large, and it had four long, narrow paddles. Unusually, the rear paddles were almost the same length as the front ones. *Temnodontosaurus* also had a large triangular dorsal fin, as did all advanced ichthyosaurs. Fossil finds of ichthyosaur young preserved inside the body of adults show that these creatures were viviparous – they gave birth to live young in the sea without having to come ashore to lay eggs.

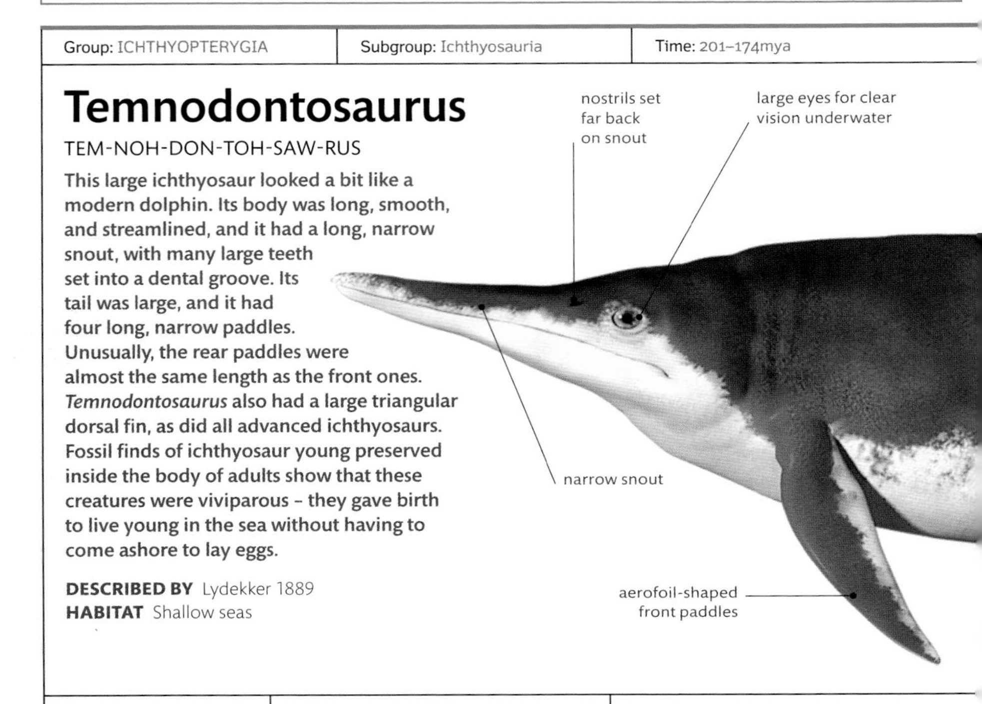

DESCRIBED BY Lydekker 1889
HABITAT Shallow seas

Length: 9m (30ft)	Weight: ~15 tonnes (14¾ tons)	Diet: Large squid and ammonites

Group: ICHTHYOPTERYGIA	Subgroup: Ichthyosauria	Time: 205–182mya

Ichthyosaurus

IK-THEE-OH-SAW-RUS

Many hundred complete skeletons of *Ichthyosaurus* have been discovered, making it one of the best-known of all prehistoric animals. It had a high dorsal fin, and broad front paddles. The end section of the tail was angled downward to support the vertical tail fin. *Ichthyosaurus* had massive ear bones, perhaps to pick up underwater vibrations created by potential prey.

DESCRIBED BY de la Beche and Conybeare 1821
HABITAT Oceans

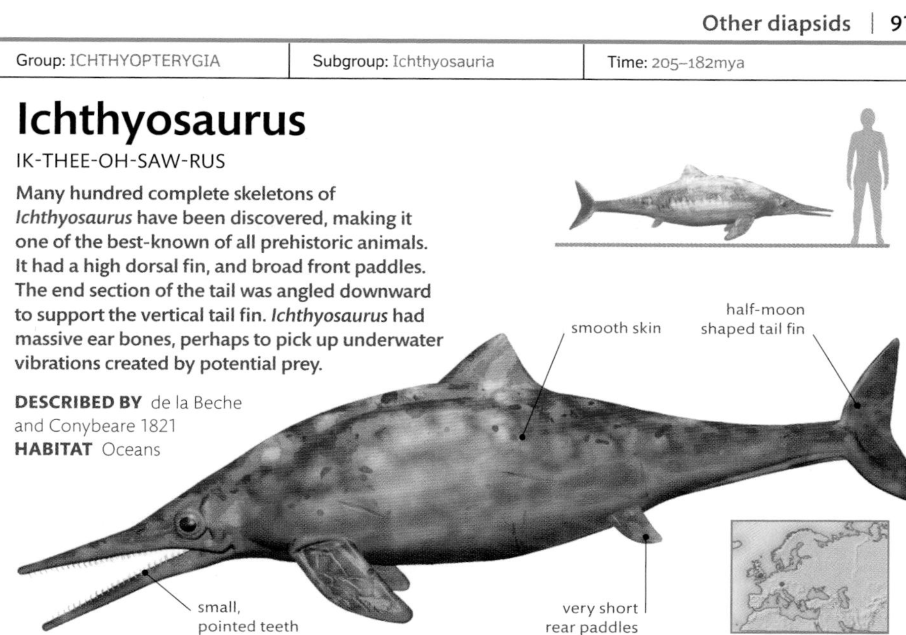

Length: 3.3m (11ft)	Weight: ~90kg (200lb)	Diet: Fish

Group: ICHTHYOPTERYGIA	Subgroup: Ichthyosauria	Time: 174–163mya

Ophthalmosaurus

OFF-THAL-MO-SAW-RUS

It is not difficult to see why *Ophthalmosaurus* ("eye lizard") was so-named. Its eyes were the largest of any creature relative to body size. As in other ichthyosaurs, the eyes were surrounded by the sclerotic ring – a circular arrangement of bony plates that prevented the soft tissues from collapsing at high pressures. Some palaeontologists think that *Ophthalmosaurus* had such large eyes because it was a night hunter – others that it swam at very deep depths. *Ophthalmosaurus* had a teardrop-shaped body. The tail had a crescent-shaped fin. The paddles were short and broad.

nostrils set high on snout, close to eyes
huge eye sockets
long, thin snout

DESCRIBED BY Seeley 1874
HABITAT Oceans

Length: 6m (20ft)	Weight: ~3 tonnes (3 tons)	Diet: Fish, squid, other molluscs

Group: PLESIOSAURIA	Subgroup: Plesiosauroidea	Time: 199–190mya

Plesiosaurus

PLEEZ-EE-OH-SAW-RUS

Plesiosaurus was built for manoeuvrability rather than speed. It had the typical plesiosaur shape of a short and wide body that tapered towards the tail. Its neck was much longer than the body, and was very flexible. Although its head was relatively small, its jaws were long, and held many sharp, conical teeth. *Plesiosaurus* had four wide paddle-shaped flippers, which it is thought to have used in a figure of eight pattern of paddling.

DESCRIBED BY de la Beche and Conybeare 1821
HABITAT Oceans

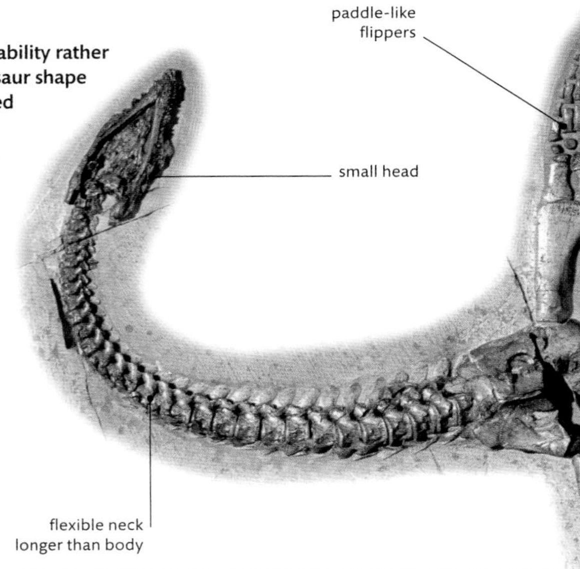

Length: 2.3m (7.6ft)	Weight: ~90kg (200lb)	Diet: Fish, squid-like molluscs

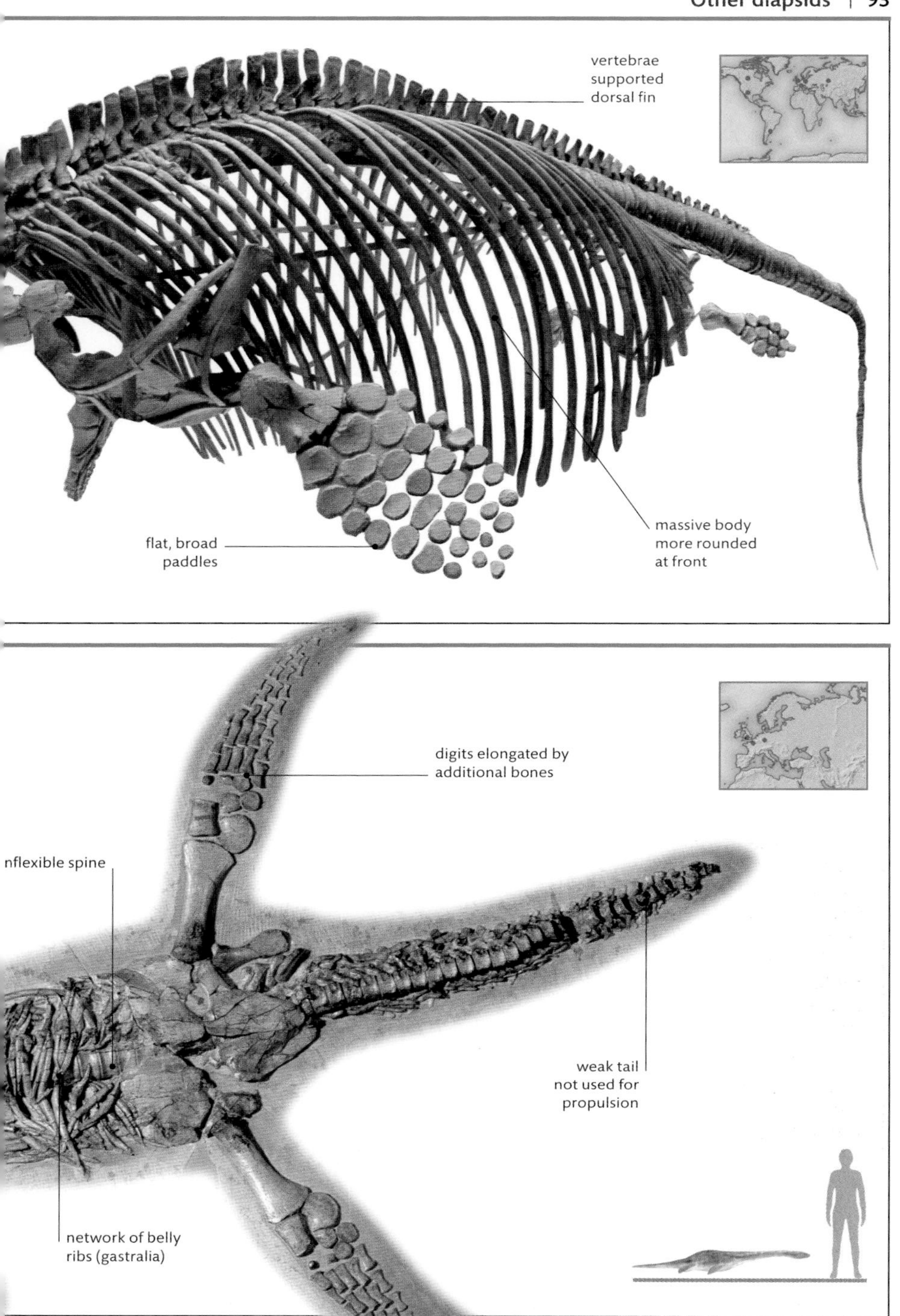
vertebrae
supported
dorsal fin
flat, broad
paddles
massive body
more rounded
at front
digits elongated by
additional bones
nflexible spine
weak tail
not used for
propulsion
network of belly
ribs (gastralia)

Group: PLESIOSAURIA	Subgroup: Plesiosauroidea	Time: 174–164mya

Cryptoclidus

CRIP-TOH-CLY-DUS

Cryptoclidus was a plesiosaur with a moderately long neck made up of 30 vertebrae. Its small skull had a long snout, with the nostrils set far back. Its jaw contained many sharp, pointed teeth. These interlocked when the jaw was closed to make a fine trap for small fish or shrimp-like prey. *Cryptoclidus* had a relatively short tail, and four hydrofoil-shaped flippers. These are thought to have been used in a figure of eight motion to propel the animal along. The shoulders and hips became large plates that supported powerful muscles that helped propel the plesiosaur. *Cryptoclidus* is thought to have given birth to live young.

DESCRIBED BY Seeley 1892
HABITAT Shallow oceans

Length: 4m (13ft)	Weight: ~8 tonnes (8 tons)	Diet: Fish, small marine animals

Group: PLESIOSAURIA	Subgroup: Pliosauroidea	Time: 166–152mya

Liopleurodon

LIE-OH-PLOOR-OH-DON

Liopleurodon was a massive marine carnivore. It had a whale-like appearance, with a large, heavy head, a short, thick neck, and a streamlined body. It is thought to have used its front and rear flippers in different swimming motions, with the front pair moving up and down, and the rear ones used in a figure of eight motion. Its teeth were arranged in a rosette at the front of its mouth. It may have swum with its mouth open, allowing water to pass into its nostrils so that it could smell its prey.

DESCRIBED BY
Sauvage 1873
HABITAT Oceans

Fossil vertebra

Length: 5–7m (16–23ft)	Weight: Not calculated	Diet: Large squid, ichthyosaurs

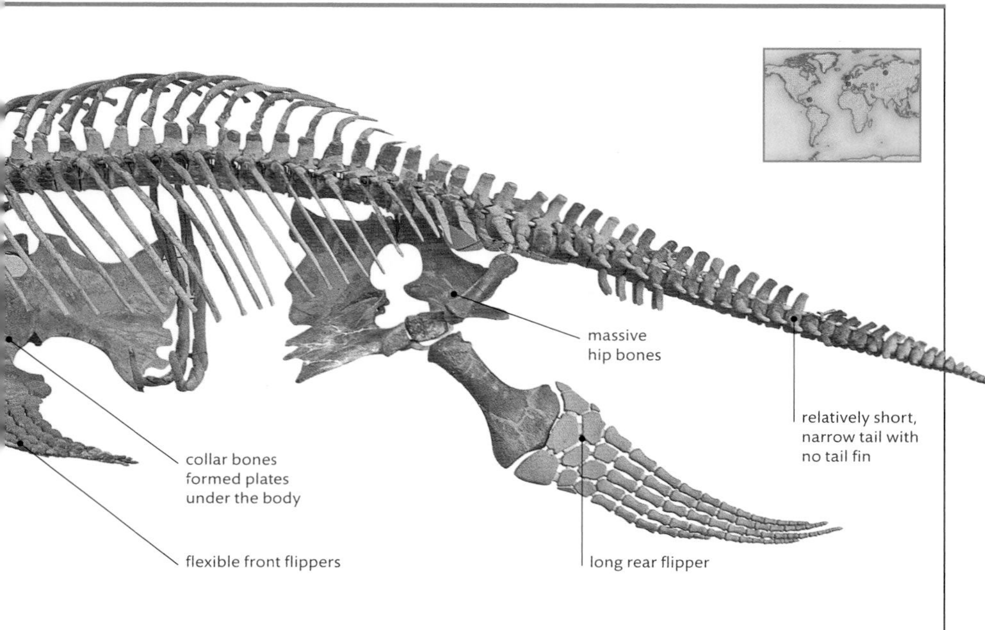
massive
hip bones
relatively short,
narrow tail with
no tail fin
collar bones
formed plates
under the body
flexible front flippers
long rear flipper

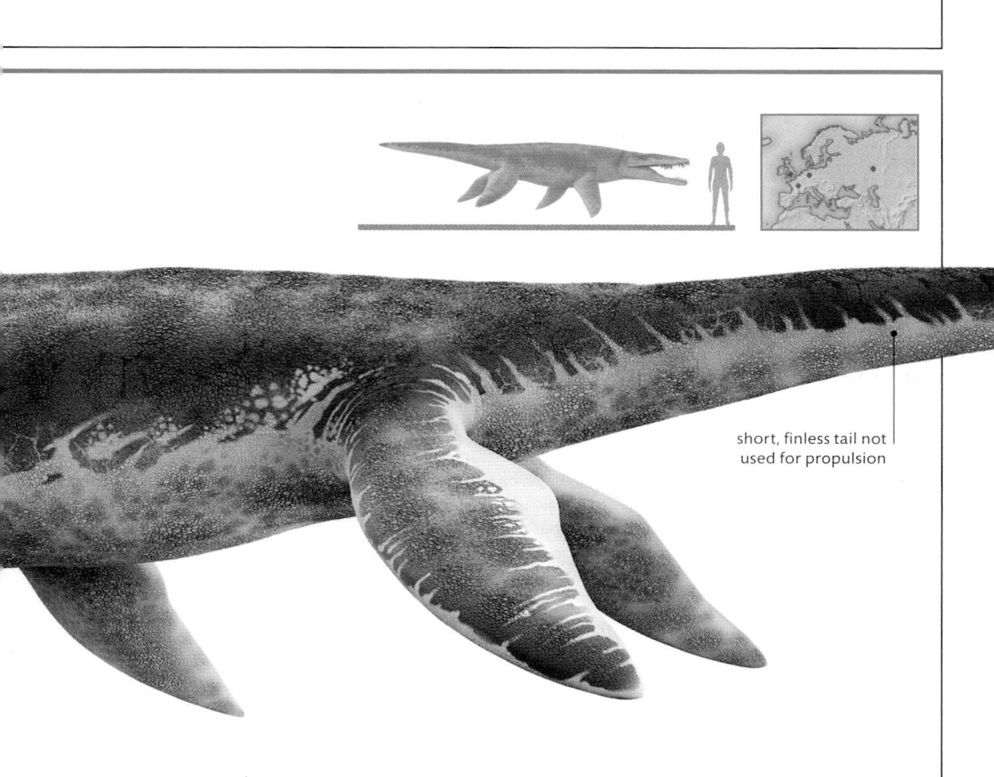
short, finless tail not
used for propulsion

Group: PSEUDOSUCHIA	Subgroup: Crocodyliformes	Time: 163–145mya

Cricosaurus

CREE-COH-SAW-RUS

This "ring lizard" was a highly specialized aquatic crocodilian. It had a streamlined shape and had lost the heavy back armour that crocodiles normally have. Its smoother skin meant that *Cricosaurus* was more manoeuvrable when swimming, and could move its body, as well as its tail, from side to side to propel itself along. It had two pairs of flippers and a large, fish-like tail fin, which was supported by a steep downward bend of the spine. One fossil specimen of *Cricosaurus* was found outlined by a thin film of carbon. This showed the shape of the limbs in life. The hind flippers were longer than the front pair, which were shaped like hydrofoils. The long, slim snout was filled with sharp, pointed teeth.

DESCRIBED BY Wagner 1858
HABITAT Seas

Length: 3m ($9\frac{3}{4}$ft)	Weight: ~120kg (265lb)	Diet: Fish

Group: PSEUDOSUCHIA	Subgroup: Crocodyliformes	Time: 166–152mya

Metriorhynchus

MET-REE-OR-RINE-CUS

An aquatic crocodilian, *Metriorhynchus* is now thought to have given birth to live young. It had a streamlined shape, although not as streamlined as its relative *Cricosaurus*, with a long, slim head, long body, and thin tail. Its skin was smooth, reducing resistance in the water, and it had a vertical, fish-like tail fin. Moving this from side to side propelled *Metriorhynchus* through the water. Its limbs were formed into paddles, with the hindlimbs being larger than the forelimbs. Its snout was long and pointed, and the jaws had long muscles to allow the animal to open its mouth very wide. The teeth were conical.

DESCRIBED BY von Meyer 1832
HABITAT Tropical seas

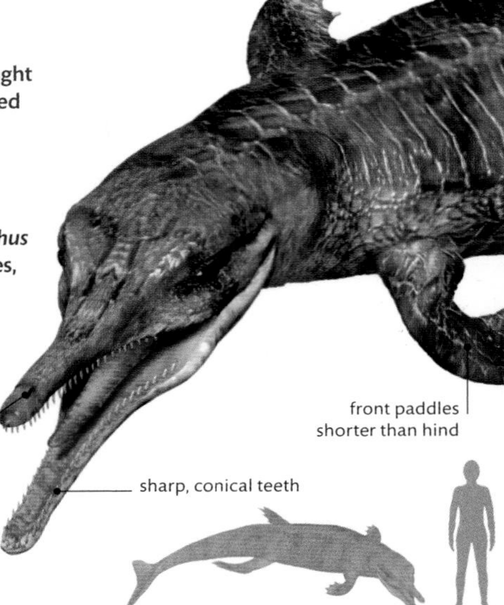

Length: 3m ($9\frac{3}{4}$ft)	Weight: ~120kg (265lb)	Diet: Fish

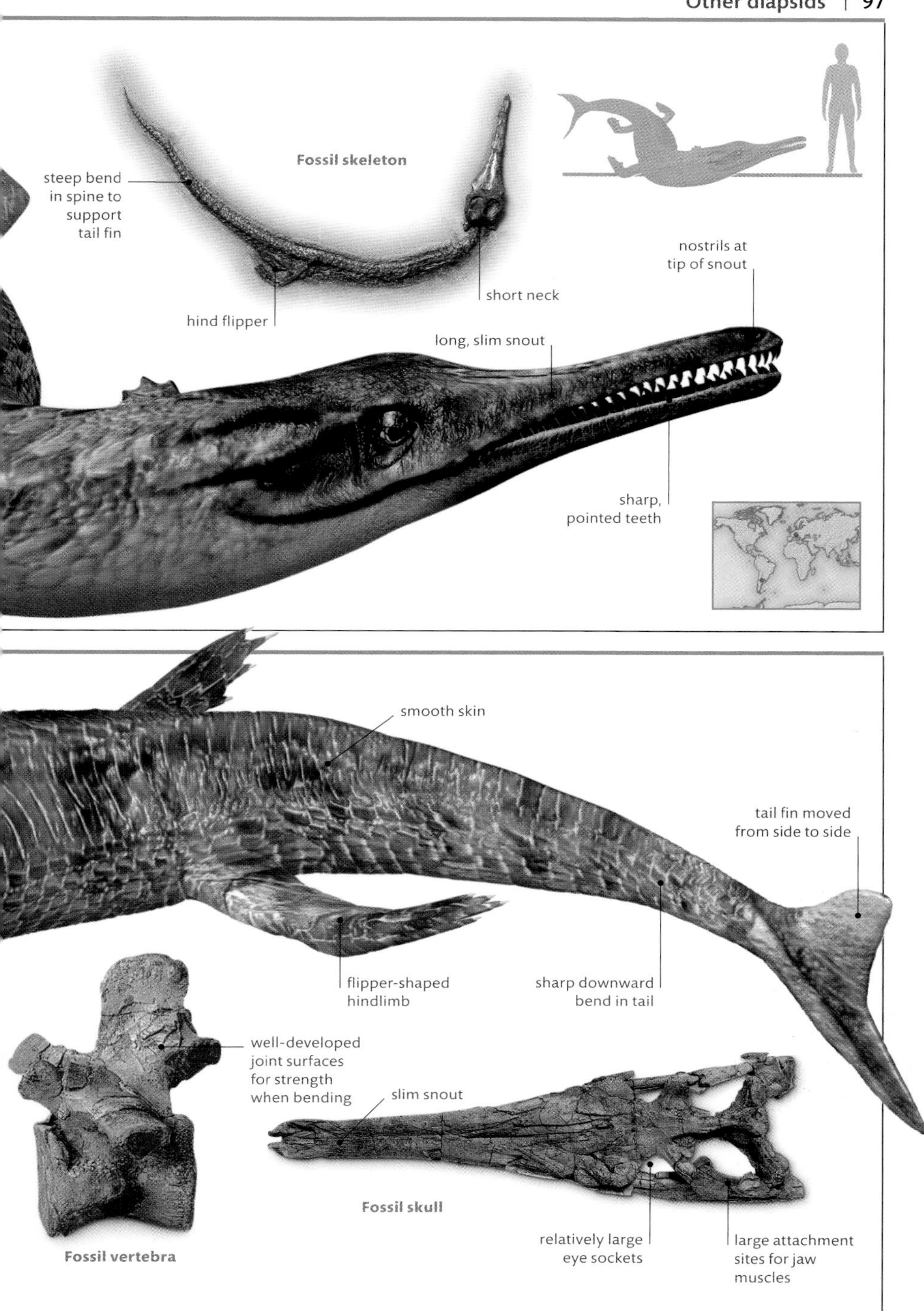
Fossil skeleton
steep bend in spine to support tail fin
hind flipper
short neck
nostrils at tip of snout
long, slim snout
sharp, pointed teeth
smooth skin
tail fin moved from side to side
flipper-shaped hindlimb
sharp downward bend in tail
well-developed joint surfaces for strength when bending
slim snout
Fossil skull
relatively large eye sockets
large attachment sites for jaw muscles
Fossil vertebra

Group: ARCHOSAURIA	Subgroup: Pterosauria	Time: 201–182mya

Dimorphodon

DIE-MOR-FOH-DON

The most striking feature of *Dimorphodon* ("two-form tooth") was its enormous puffin-like head. It had a short neck and a long tail. The tail could only be moved near the base, helping to steer the animal. It was probably a clumsy walker. *Dimorphodon* was a relatively poor flier that is suggested to have spent most of its time hanging from cliffs or branches. It is thought to have been a quadrupedal climbing animal that used its short wings for short flights only as a last resort. *Dimorphodon* had two types of teeth – long front ones and small cheek teeth.

DESCRIBED BY Owen 1859
HABITAT Shores

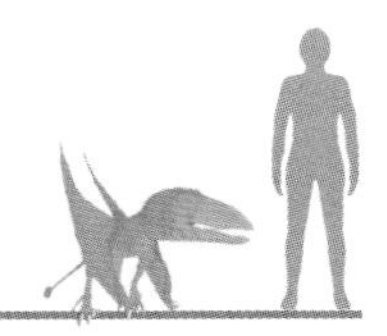

long front teeth

Length: 1.4m (4ft)	Weight: Not calculated	Diet: Fish

Group: ARCHOSAURIA	Subgroup: Pterosauria	Time: 150–148mya

Rhamphorhynchus

RAM-FOR-RINK-US

This pterosaur is well-known from beautifully preserved specimens found along with *Archaeopteryx*. Details of the wing structure and the tail can be seen, and a throat pouch has even been preserved in a few fossils. *Rhamphorhynchus* had long narrow jaws to act as a fish spear. Its sharp teeth pointed outwards. It had a long tail with a diamond-shaped flap of skin at the end, and small legs. The wing membrane stretched down to the ankle.

DESCRIBED BY von Meyer 1847
HABITAT Shores

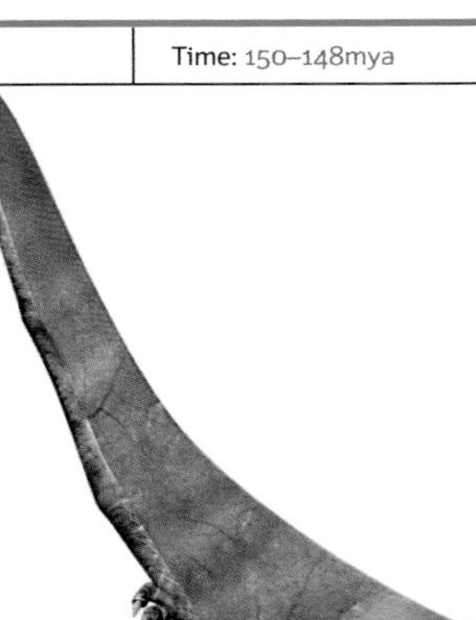

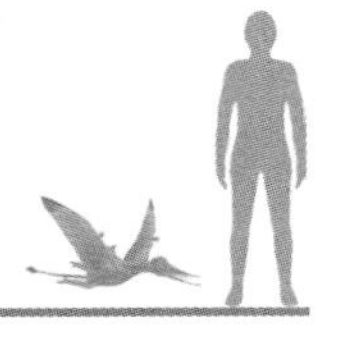

large eyes

outward-pointing teeth at end of jaws

Length: 1m (3¼ft)	Weight: ~20kg (44lb)	Diet: Fish

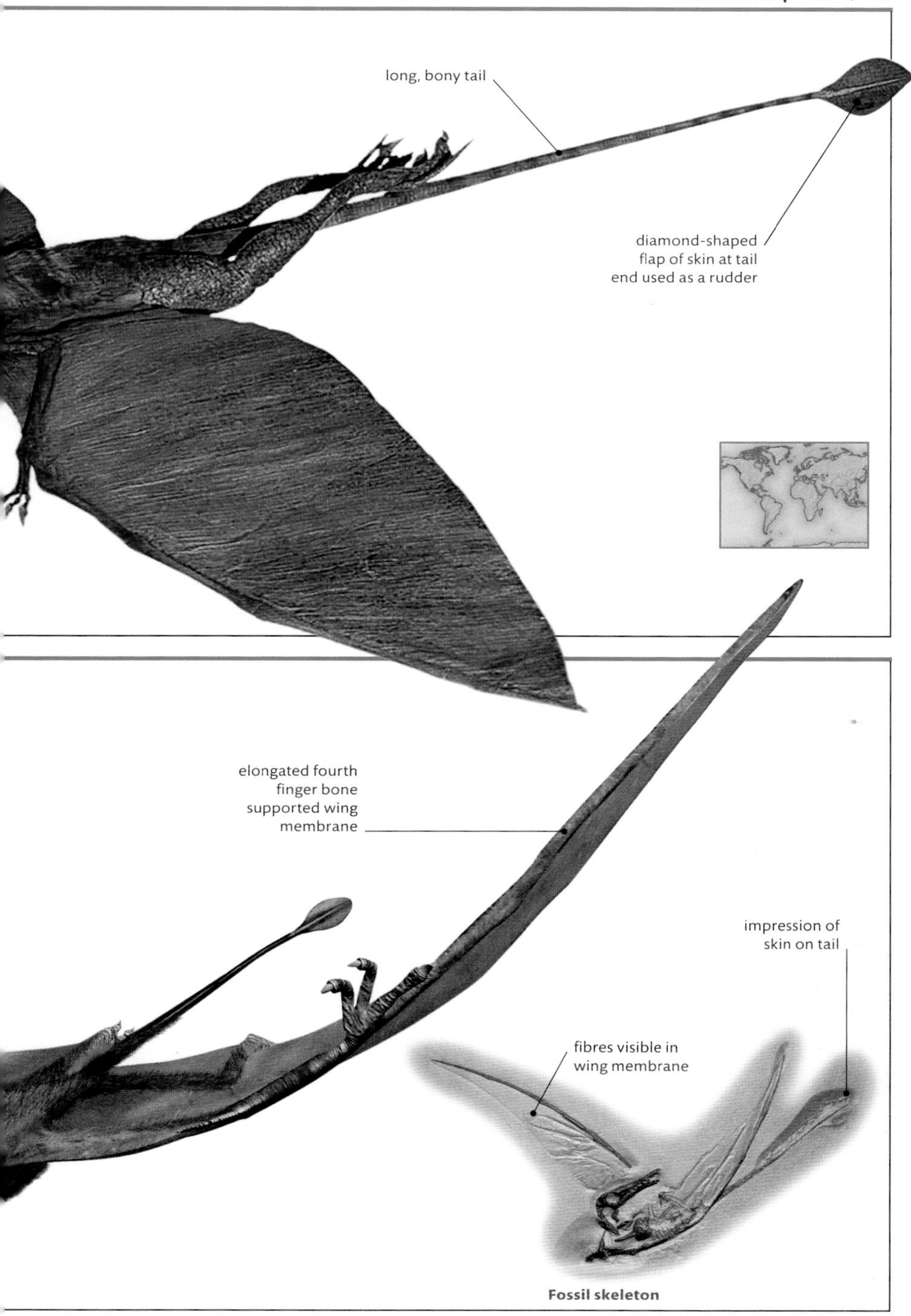
long, bony tail
diamond-shaped flap of skin at tail end used as a rudder
elongated fourth finger bone supported wing membrane
impression of skin on tail
fibres visible in wing membrane

Fossil skeleton

Group: ARCHOSAURIA	Subgroup: Pterosauria	Time: 150–148mya

Anurognathus

AN-YOOR-OG-NATH-US

This tiny pterosaur was unlike other rhamphorhynchoids in that it only had a very short tail. In fact, its name means "without tail and jaw". Along with the short tail, its small body size would have made manoeuvring on the wing very easy. *Anurognathus* is now known to have a very short, broad skull, similar to a modern nightjar. The peg-like teeth were used for catching insects. Its wings, which reached a maximum span of about 50cm (20in), were composed of a thin flap of skin that stretched from the elongated fourth finger to the ankle. A secondary wing, supported by the pteroid bone, stretched from each wrist to the neck.

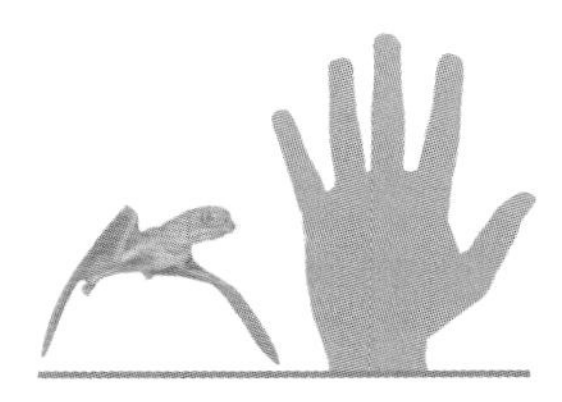

DESCRIBED BY Döderlein 1923
HABITAT Wooded plains

thin, delicate wing membrane

Length: 9cm (3½in)	Weight: ~7g (¼oz)	Diet: Insects

Group: PTEROSAURIA	Subgroup: Pterodactyloidea	Time: 150–148mya

Pterodactylus

TER-OH-DAK-TIL-US

Of the many species of *Pterodactylus* ("wing finger") originally discovered, most have now been grouped into a single species while those that remain separate have been moved to other genera. Only one species of *Pterodactylus* is currently recognized. It had a long, curved, pelican-like neck, and a long skull with many small, pointed teeth. Skeletal analysis shows that it was a powerful, active flyer.

DESCRIBED BY Cuvier 1809
HABITAT Shores

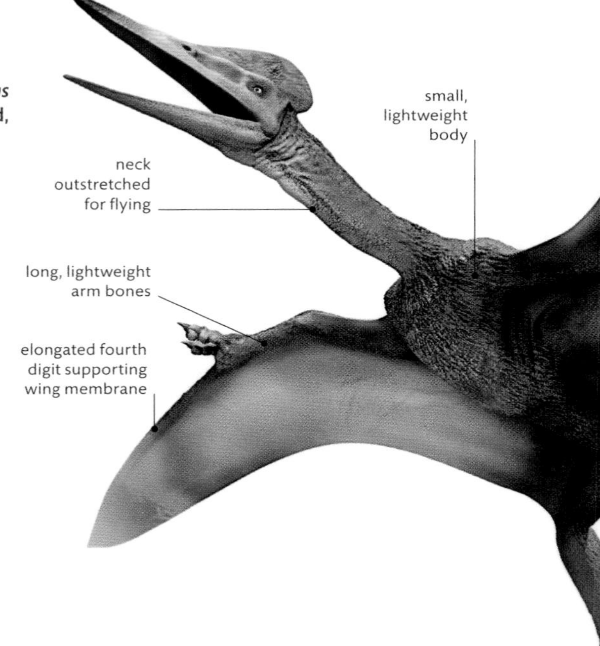

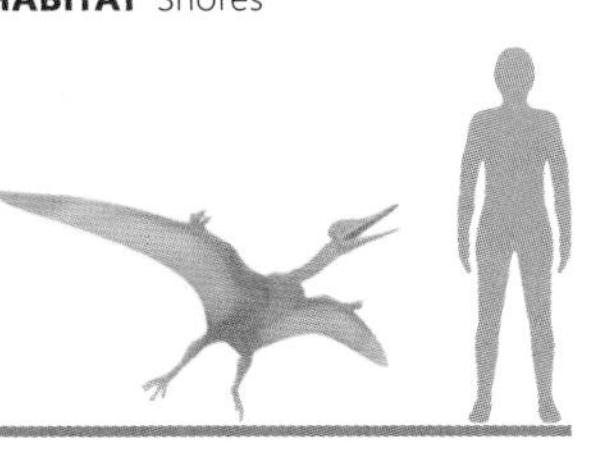

Length: 1m (3¼ft)	Weight: ~1–5kg (2–11lb)	Diet: Fish

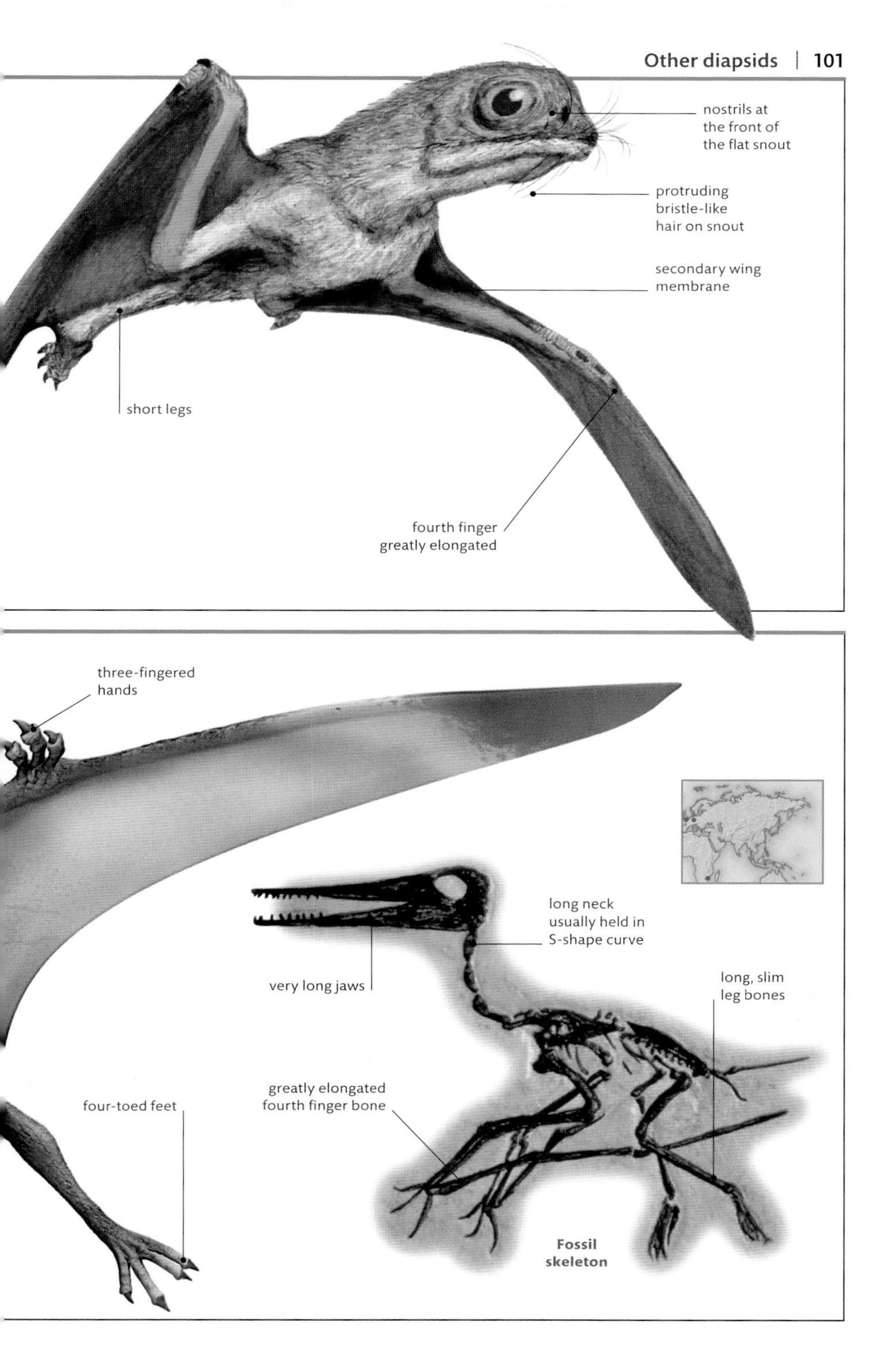
nostrils at
the front of
the flat snout
protruding
bristle-like
hair on snout
secondary wing
membrane
short legs
fourth finger
greatly elongated
three-fingered
hands
long neck
usually held in
S-shape curve
very long jaws
long, slim
leg bones
greatly elongated
fourth finger bone
four-toed feet

Fossil skeleton

SYNAPSIDS

ONLY THE CYNODONTS survived into the Jurassic, and even then, only a select few lineages.

For a long time, we knew very little about these "proto-mammals", called Mammaliaformes, because their fossil record consists mostly of teeth and rare jaws. It was therefore believed that throughout the Jurassic and most of the Cretaceous, mammals and their close relatives were struggling to survive in the shadows of the dinosaurs.

In the last 25 years, this view has been radically changed by a trove of incredible fossils from the Jurassic found in China. These beds preserve some of the oldest true mammals, and they show that three lineages - docodonts, haramiyidans, and multituberculates - were incredibly diverse. These groups include not only nocturnal insectivores, but swimmers, climbers, diggers, and even gliders.

These fossils also record the gradual accumulation of the features we now consider characteristic of mammals, including specialized teeth, with a complex topography of cusps, basins, and crests. To stabilize these new teeth for precise chewing motions, the tooth-bearing jawbone, called the dentary, became anchored to the skull, and the smaller jawbones behind it were reduced. These small bones were eventually incorporated into the hearing apparatus, part of a series of changes that heightened the senses of mammals. These adaptations allowed early mammals to acquire the energy needed to drive their fast metabolisms.

Group: MAMMALIAFORMES	Subgroup: Docodonta	Time: 156–149mya

Docodon

DOH-CO-DON

Docodon ("beam tooth") was a mouse-sized, close mammal relative. So far, this genus is known only from remains of jaws and teeth. These had very complex tooth cusp patterns, allowing *Docodon* to chew food effectively. It appears to have been a rodent-like animal, and was probably endothermic.

DESCRIBED BY
Marsh 1881

HABITAT
Forests

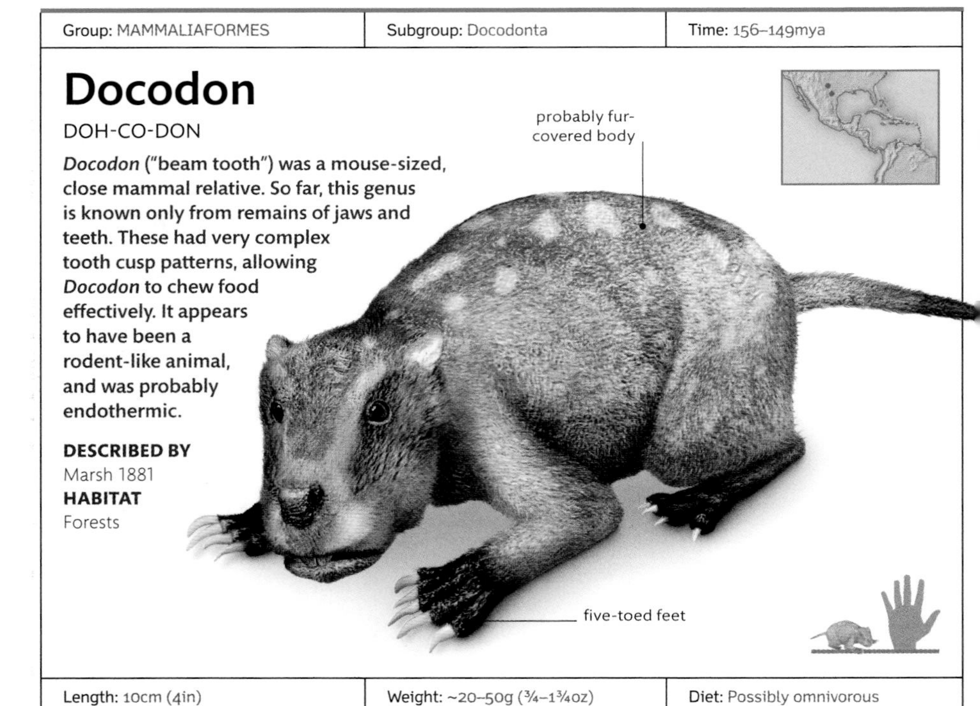

Length: 10cm (4in)	Weight: ~20–50g (¾–1¾oz)	Diet: Possibly omnivorous

Group: CYNODONTIA	Subgroup: Mammaliaformes	Time: 199–190mya

Sinoconodon

SINE-OH-CO-DON

Sinoconodon ("Chinese spiky tooth") was a small, primitive mammal resembling a modern shrew. Its skull structure, despite retaining some early cynodont characters, showed that it was very closely related to true mammals. In particular, the structures housing the inner ear are similar to those of true mammals. Although the cheek-teeth were permanent, as in living mammals, its other teeth were still replaced several times during its lifetime, like in earlier cynodonts. The rear of the braincase had expanded and the eye socket was large. *Sinoconodon* was apparently endothermic, and was probably covered in fur. It was a quadruped, with five long, clawed toes on each foot, and a long tail. The snout was long and slim.

DESCRIBED BY Patterson and Olson 1961
HABITAT Woodland

fur-covered body
long, flexible tail
relatively large braincase
large eyes
slim, pointed snout with mammalian teeth
five clawed toes

Length: 10–15cm (4–6in)	Weight: ~30–80g (1–2⅞oz)	Diet: Probably mainly insects

Group: MAMMALIAFORMES	Subgroup: Mammalia	Time: 152–139mya

Triconodon

TRY-CON-OH-DON

Triconodon ("three-spiked tooth") was a small raccoon- or possum-like early mammal. Originally classed in the same group as *Eozostrodon* (see p.57), recent analysis shows that it was a true mammal. Its teeth were rather unspecialized and versatile, suggesting an omnivorous lifestyle, including insects and small reptiles. The development of a mammalian middle ear seems to have coincided with an increase in brain size.

DESCRIBED BY Owen 1859
HABITAT Woodland

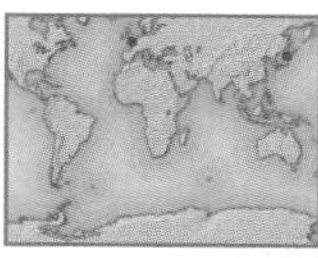

molars with three sharp points
large teeth at the front of the jaw

Fossil jawbone

Length: 50cm (20in)	Weight: ~750g (1¾lb)	Diet: Omnivorous

CRETACEOUS PERIOD

145–66 MILLION YEARS AGO

THE CLIMATE in the Cretaceous was similar to that in the Jurassic – warm and humid. Sea levels were very high, resulting in large areas of flooding and consequent isolation of landmasses. A major evolutionary development – the diversification of the angiosperms (flowering plants) led to important changes in the landscape and animal life on Earth. Many new insect groups appeared and dinosaurs continued to flourish. Populations of certain groups, such as sauropods and stegosaurs, were greatly reduced as their staple food supply was taken over by new dinosaur groups, including the advanced ornithopods and ceratopsians. These flourished until the mass extinction event at the end of the period.

Flowering plants
Late in the Cretaceous, forests were composed of flowering trees, such as birch and magnolia. *Betulites* (above) is an extinct birch species.

4,600mya 4,000mya 3,000mya

CRETACEOUS LANDMASSES

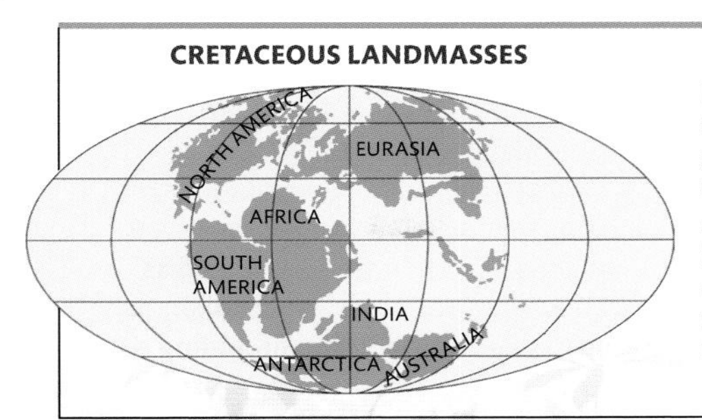

During the Cretaceous, continental drift led to the break up of the supercontinents of Laurasia and Gondwana. This process was augmented by the very high sea levels of 200m (640ft). The Atlantic Ocean started to open up between the future South America and Africa, and Antarctica was almost in its modern position. India started to move northwards. There was still a land link between Eurasia and North America.

CRETACEOUS LIFE

Dinosaurs were still the dominant land animals. They increased in diversity throughout the period, with new groups including the hadrosaurs, horned dinosaurs, and giant tyrannosaurs appearing. Birds continued to evolve and snakes appeared. Mammals were still small, but continued to diversify.

Mammal evolution
Mammals remained only minor elements of the fauna. Like the early marsupial *Didelphodon* (right) they were shrew- or cat-sized animals.

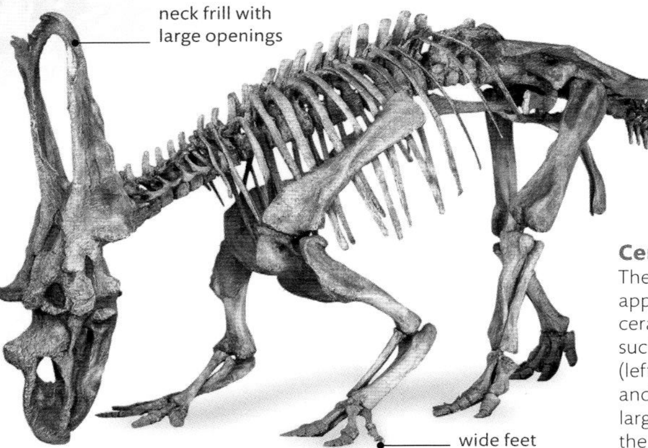

Ceratopsians
The Cretaceous saw the appearance of the frilled ceratopsian dinosaurs, such as *Chasmosaurus* (left). These had brow and nose horns, and a large, bony frill covering the neck and shoulders.

THEROPODS

THEROPODS REACHED their peak in diversity in the Cretaceous. Ceratosaurs and carnosaurs remained the top predators in the Early Cretaceous, and some, such as *Giganotosaurus* and *Spinosaurus*, reached enormous sizes in the early parts of the Late Cretaceous.

During the mid-Cretaceous, ecosystems underwent drastic changes alongside the spread of flowering plants. Some coelurosaurian theropods became larger and started to displace the ruling tetanuran theropods. In Laurasia, the ceratosaurs and carnosaurs became extinct, possibly because of the decline of certain sauropods, but in Gondwana, these groups persisted with coelurosaurs until the end of the Cretaceous. Tyrannosaurids filled the void of Late Cretaceous apex predators in Laurasia, spreading throughout Asia and North America.

Some coelurosaur groups, such as the sickle-clawed dromaeosaurs and troodontids became highly specialized. While others, such as therizinosaurs, oviraptorosaurs, and ornithomimosaurs became herbivorous, losing teeth and developing a keratinous beak instead.

During the Early Cretaceous, toothed enantiornithine birds were common, but toothless ornithurine birds became predominant by Late Cretaceous. It was debated whether modern bird lineages arose before or after the extinction, but new fossils now conclusively show that they originated in the latest Cretaceous.

Group: CERATOSAURIA	Subgroup: Abelisauridae	Time: 83–78mya

Abelisaurus

AH-BEL-EE-SAW-RUS

This large theropod, named for its discoverer, Roberto Abel, is known only from a single skull. It seems to have been rather a primitive form related to *Carnotaurus*. Its head was large, with a rounded snout, and its teeth were relatively small for a carnivorous dinosaur of its size. Its skull is peculiar in that it has a huge opening at the side just above the jaws, which is much larger than in other dinosaurs. Reconstructions of the rest of its body are necessarily hypothetical. The illustration here is based on the body structure of *Carnotaurus* and similar theropod dinosaurs.

DESCRIBED BY Bonaparte 1984
HABITAT River plains

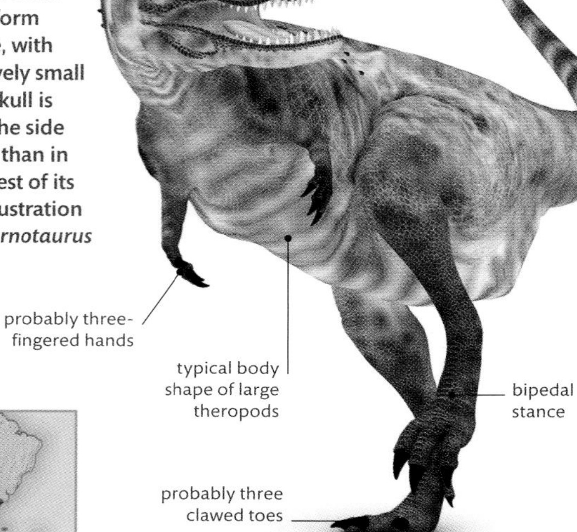

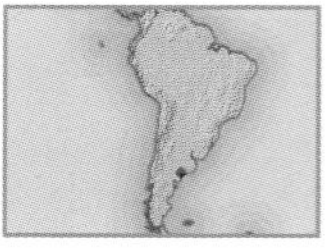

Length: 9m (30ft)	Weight: ~1.4 tonnes (1³⁄₈ tons)	Diet: Meat

Group: CERATOSAURIA	Subgroup: Abelisauridae	Time: 82–66mya

Carnotaurus

KAR-NOH-TAW-RUS

The most distinctive feature of *Carnotaurus* ("meat-eating bull") were the broad, triangular horns over its eyes. It did not look like other large theropods, and had a short, tall skull, tiny arms with reduced hands, and long legs. Its lightly built body was covered in scales and studs. Its three fingers were short and stubby, and the thumbs bore a small spike. Its tail was long, thick, and flexible.

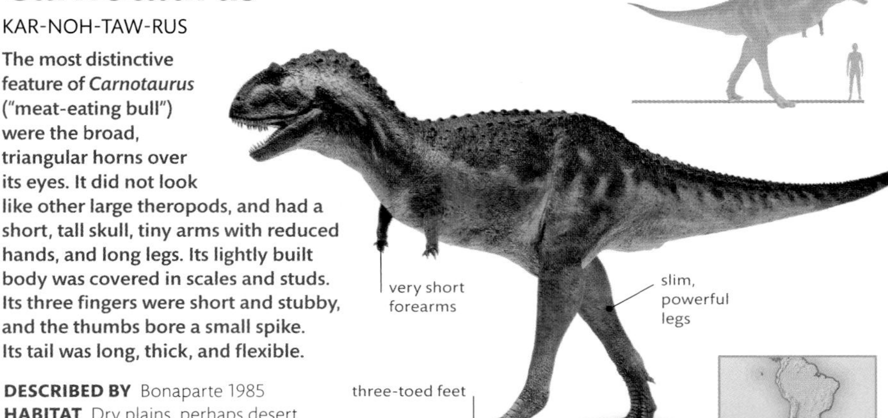

DESCRIBED BY Bonaparte 1985
HABITAT Dry plains, perhaps desert

Length: 7.5m (25ft)	Weight: ~1 tonne (1 ton)	Diet: Meat

Group: CARNOSAURIA	Subgroup: Carcharodontosauridae	Time: 100–96mya

Giganotosaurus

GIG-AN-OH-TOE-SAW-RUS

***Giganotosaurus* ("giant southern lizard") is one of the largest carnivorous dinosaurs yet discovered. Its skull was longer than an average man, and held long, serrated teeth. Its hands had three fingers, and it had a slim, pointed tail. Although larger than *Tyrannosaurus*, it was more lightly built than that dinosaur, and seems to have hunted in a different fashion – by slashing at its prey rather than charging and biting head on. Despite its huge size, some paleontologists maintain that, like other large theropods, it may have been able to run relatively fast.**

large eyes

serrated teeth 20cm (8in) long

three clawed fingers

DESCRIBED BY Coria and Salgado 1995
HABITAT Warm swamps

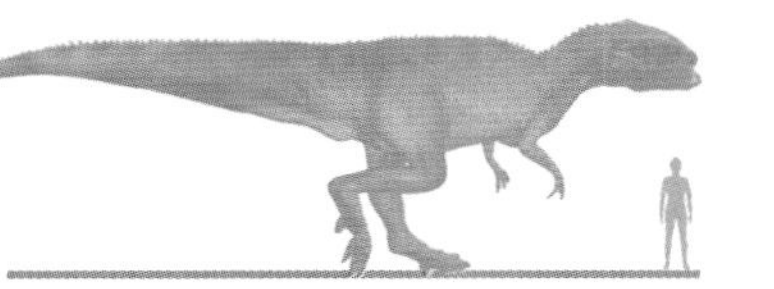

Length: 12.5m (41ft)	Weight: ~8 tonnes (7⅞ tons)	Diet: Meat

Group: CARNOSAURIA	Subgroup: Spinosauridae	Time: 98–94mya

Spinosaurus

SPINE-OH-SAW-RUS

***Spinosaurus* ("spine lizard") was an immense theropod with a crocodile-like snout and a sail running all the way down its back. It was among the largest carnivores to ever walk the Earth, rivalling *Giganotosaurus* and *Tyrannosaurus* in size, but its exact length remains uncertain. New discoveries show that the hindlimbs were short and the tail was tall and paddle-like, which may have been adapted to suit a semi-aquatic lifestyle. However, a number of palaeontologists have argued that some of the fossils are from a different, closely-related spinosaur, and that the sail on its back would have made it unstable while swimming.**

DESCRIBED BY Stromer 1915
HABITAT Tropical swamps

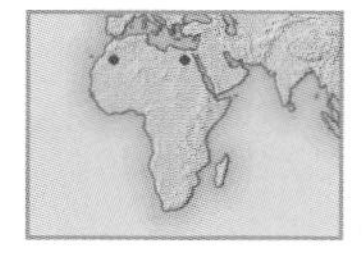

tooth socket

Fossil tooth battery

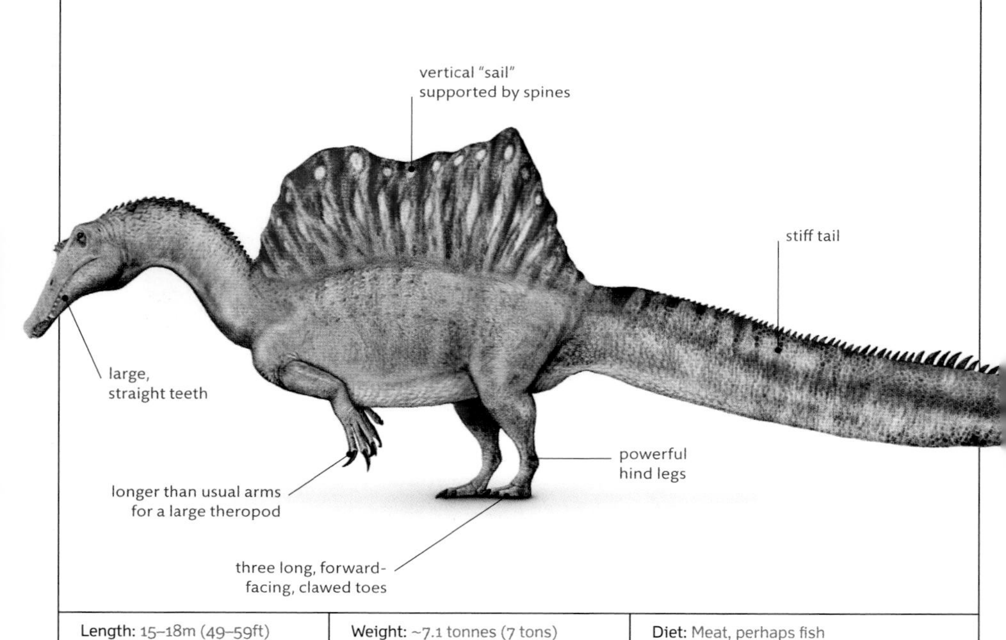

Length: 15–18m (49–59ft)	Weight: ~7.1 tonnes (7 tons)	Diet: Meat, perhaps fish

Group: CARNOSAURIA	Subgroup: Spinosauridae	Time: 129–125mya

Suchomimus

SOOK-OH-MIME-US

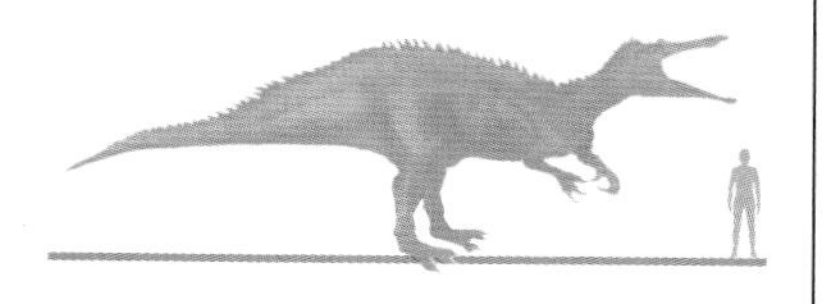

This dinosaur, whose name means "crocodile-mimic", had several features that indicate that it ate fish. It had a very long snout, and huge, curved thumb claws that could be used for hooking fish from the water. The jaws had over 100 teeth that pointed slightly backwards, and the end of the snout had a rosette of longer teeth. A low ridge ran along the length of its back. Its arms were relatively long, which would have enabled *Suchomimus* to reach into the water to grasp prey.

DESCRIBED BY Sereno et al. 1998
HABITAT Lush forests

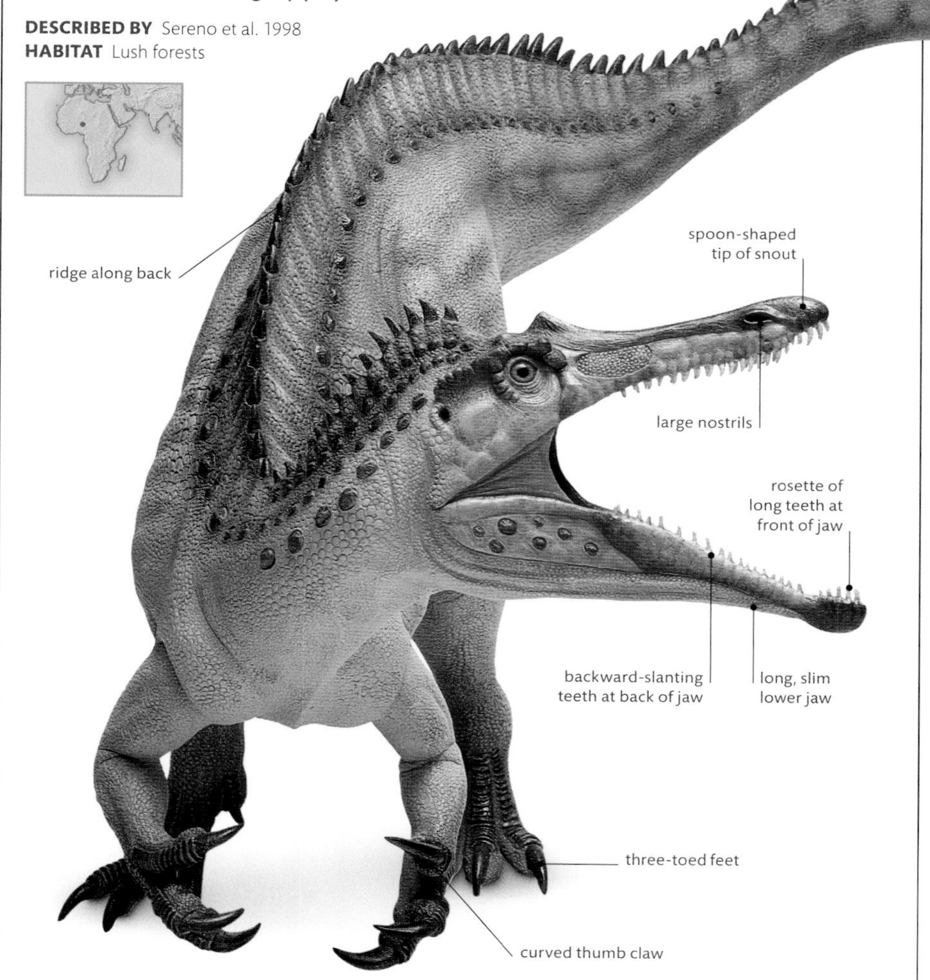

Length: 11m (36ft)	Weight: ~5.1 tonnes (5 tons)	Diet: Fish, possibly meat

Group: CARNOSAURIA	Subgroup: Spinosauridae	Time: 129–125mya

Baryonyx

BAR-EE-ON-ICKS

Baryonyx ("heavy claw") was named for one of its unusual features – its huge, curved thumb claws. Its skull was also an unusual shape for a theropod, being long and narrow – rather like that of a crocodile. There was a bony crest on the top of the head, and the jaws were filled with 96 pointed and serrated teeth. This is twice as many as theropods usually possessed. Its arms were thick and unusually powerful. These features, along with the remains of fish found with the skeleton, have led palaeontologists to surmise that *Baryonyx* fished using its claws as hooks, in the same way as a modern bear does.

DESCRIBED BY Charig and Milner 1986

HABITAT Riverbanks

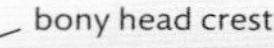

BARYONYX CLAW

Although the thumb claws were not found attached to the skeleton, they were initially suspected to have been from the foot. However, recent finds have now made it clear that they formed part of the hand. Palaeontologists reconstructed *Baryonyx* with the claws in this position because the length and thickness of the arms seemed to match their proportions.

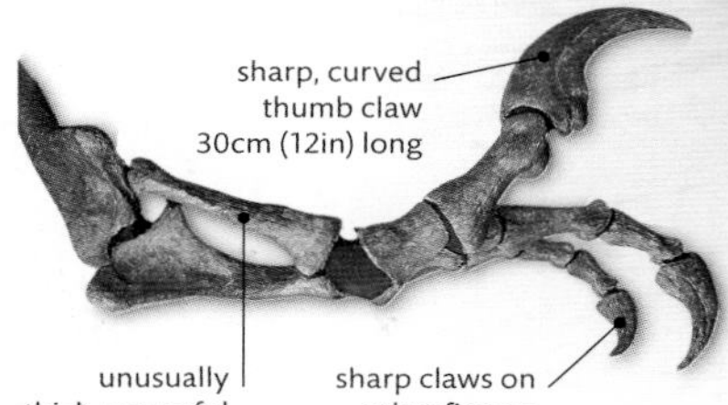

Length: 10m (33ft)	Weight: ~2 tonnes (2 tons)	Diet: Fish, perhaps meat

Baryonyx skull

Group: COELUROSAURIA	Subgroup: Tyrannosauridae	Time: 68–66mya

Tyrannosaurus

TIE-RAN-OH-SAW-RUS

Tyrannosaurus ("tyrant lizard") was one of the largest terrestrial carnivores ever. It was a heavily-built theropod, with thick, long, powerful legs and a large, deep head. Palaeontologists debated whether it was an active predator or simply a scavenger, and how fast it actually was. However, recent analysis shows that *Tyrannosaurus* was a predator that would opportunistically scavenge. Evidence supporting this includes numerous healed bites from tyrannosaurs found on hadrosaur and ceratopsian bones, showing that it was hunting them while they were alive. For most of its lifetime, *Tyrannosaurus* would have been able to run as quickly as its prey, but eventually it was restricted to a walking gait due to its body size. The presence of feathers on its body is still debated, but skin impressions show that most of the animal was covered in pebbly scales. If feathers were present, they were limited to the back.

DESCRIBED BY Osborn 1905
HABITAT Open forests, coastal forested swamps

New discovery
In 1990 a new specimen of *Tyrannosaurus* was discovered in South Dakota, USA. The skeleton is called Sue, after its discoverer Sue Hendrickson, and is 90 per cent complete.

Length: 12m (40ft)	Weight: ~6.7 tonnes (6⅝ tons)	Diet: Hadrosaurs, ceratopsians

TYRANNOSAURUS SKELETON

The first skeletal reconstruction of *Tyrannosaurus* was prepared in 1915, showing the dinosaur standing upright with its tail along the ground. Recent analysis has shown that the backbone was held horizontally, with the body perfectly balanced at the hips.

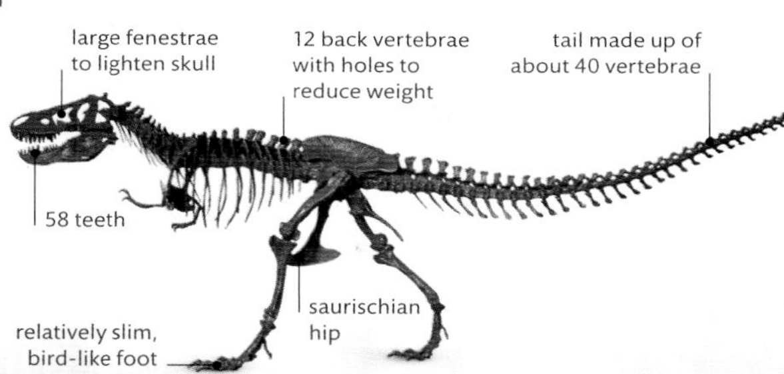

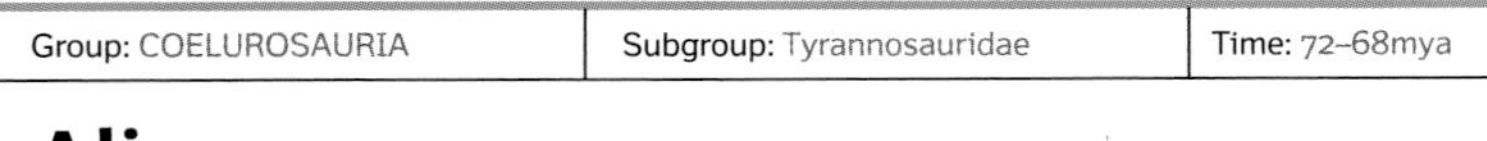

Group: COELUROSAURIA	Subgroup: Tyrannosauridae	Time: 72–68mya

Alioramus

AL-EE-OH-RAY-MUS

Alioramus ("different branch") was a small relative of *Tyrannosaurus rex* with a long, slender snout. Adorning the snout and skull were multiple small horns and bumps, that may have been used in mating displays. The lighter build of *Alioramus* compared to its larger cousins likely reflected their different prey preferences. *Alioramus* probably ate small animals instead of bigger ceratopsians and hadrosaurs.

elongated snout with small horns and bumps

short arms, like other tyrannosaurs

long, agile legs

DESCRIBED BY Kurzanov 1976
HABITAT Woodland

Length: 5m (16ft)	Weight: ~369kg (813½lb)	Diet: Meat

Group: COELUROSAURIA	Subgroup: Compsognathidae	Time: 124–122mya

Sinosauropteryx

SIEN-OH-SORE-OP-TUH-RIKS

Sinosauropteryx ("Chinese lizard wing") was the first dinosaur discovered with feather impressions. These feathers were simpler than those of living birds: they looked more like filaments. By studying the pigment-holding structures of these feathers, palaeontologists have learned that *Sinosauropteryx* was reddish-brown, with a striped tail. Its countershaded body would have allowed it to camouflage effectively in the open environments it inhabited.

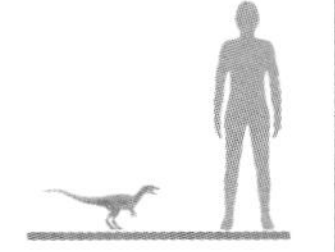

fuzzy feathers

striped tail for camouflage

relatively short arms

three clawed toes

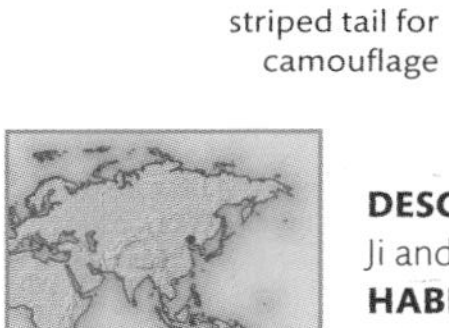

DESCRIBED BY
Ji and Ji 1996
HABITAT
Open forests

Length: 1m (3ft)	Weight: ~500g (18oz)	Diet: Lizards and small animals

Group: COELUROSAURIA	Subgroup: Ornithomimidae	Time: 72–68mya

Gallimimus

GAL-EE-MY-MUS

Gallimimus ("chicken mimic") is one of the best-known of the ornithomimids or "bird-mimic" dinosaurs. It had a short body with a long, stiff tail, and slim legs built for running at high speed. Its neck was slender and flexible, and the skull ended in a long, toothless beak. The braincase was relatively large, and *Gallimimus* was probably quite intelligent. Its eyes were large, but it did not have stereoscopic vision.

DESCRIBED BY Osmólska et al. 1972
HABITAT Desert plains

large eye socket

toothless beak

Fossil skull

slender feet with three toes

long, grasping arms

Length: 6m (20ft)	Weight: ~400kg (880lb)	Diet: Omnivorous

Group: COELUROSAURIA	Subgroup: Ornithomimosauria	Time: 72–68mya

Deinocheirus

DINE-OH-KIRE-US

The first remains of *Deinocheirus* were an enormous pair of arms 2.4m (8ft) long, which gave rise to the dinosaur's name – "terrible hand". New specimens show that *Deinocheirus* was one of the ornithomimosaurs, or bird-mimic dinosaurs. It had a long, low skull with a deep lower jaw and a duck beak. There was a hump on its back above its hips, its hindlimbs were short, and its feet were broad with hoof-like claws. Gastroliths and fish remains were preserved in its stomach, which suggests that it was an omnivore.

body and tail probably similar to other theropods

three fingers on each hand

25cm- (10in-) long claws

DESCRIBED BY
Kielan-Jaworowska 1969
HABITAT
Desert

Length: 11m (36ft)	Weight: ~6,300 kg (6.9 tons)	Diet: Omnivore

Group: COELUROSAURIA	Subgroup: Ornithomimidae	Time: 77–76mya

Struthiomimus

STRUTH-EE-OH-MY-MUS

For many years after its first discovery, *Struthiomimus* ("ostrich mimic") was thought to be the same dinosaur as *Ornithomimus* (below). The two dinosaurs are remarkably similar - the main difference is that *Struthiomimus* had longer arms with stronger fingers. In addition, its thumbs did not oppose the fingers. This reduced the grasping ability of the hands. *Struthiomimus*, like all of the other "bird-mimics", had a small head with a toothless beak and a long, stiff tail.

small head

strong fingers

long foot bones

DESCRIBED BY Osborn 1916
HABITAT Open country, riverbanks

Length: 3.5m (11ft)	Weight: ~150kg (330lb)	Diet: Omnivorous

Group: COELUROSAURIA	Subgroup: Ornithomimidae	Time: 77–67mya

Ornithomimus

OR-NITH-OH-MY-MUS

This "bird mimic" is typical of its family. It had slim arms and long, slim legs. Its tail made up more than half of its length, and was kept stiff by a network of ligaments. Its small head had a toothless beak, and was held upright by an S-shaped bend in the long, flexible neck. The brain cavity was relatively large. *Ornithomimus* ran with its body held horizontally and its tail outstretched for balance. Its top speed has been estimated at up to 30mph (50kph).

Fossil footprints

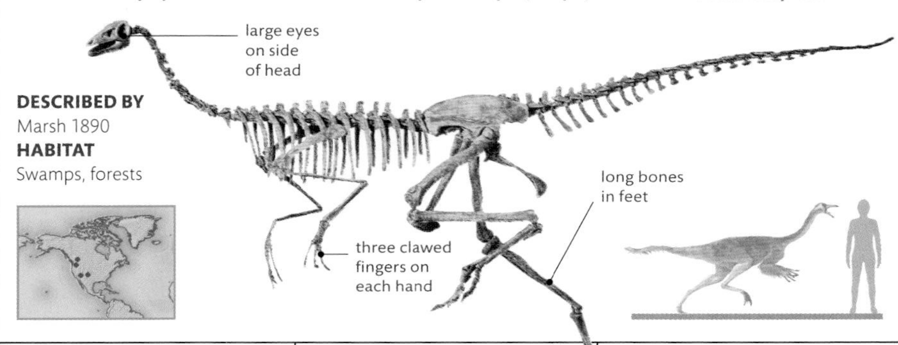

DESCRIBED BY
Marsh 1890
HABITAT
Swamps, forests

Length: 3.5m (11ft)	Weight: ~175kg (385lb)	Diet: Omnivorous

Group: COELUROSAURIA	Subgroup: Alvarezsauridae	Time: 83–76mya

Shuvuuia

SHU-VOO-EE-A

Since its discovery, *Shuvuuia* (its name is derived from "*shuvuu*", the Mongolian word for bird) has provoked great debate as to whether it was a dinosaur-like bird, or a bird-like dinosaur. It is now clear that *Shuvuuia* and its kin were not birds, but they had converged on some of the distinctive features of modern birds. Although *Shuvuuia* did not have a true beak, its slender jaws bore tiny teeth, and it may have been able to raise its upper jaw in relation to the braincase, a facility only modern birds have. Chemical analysis also shows that it was feathered. The legs were long and slim, suggesting that it may have been a fast runner. The forelimbs were short and stubby, and ended in a single, clawed digit. These unusual arms were probably used for digging at insect nests.

DESCRIBED BY Chiappe et al. 1998
HABITAT Woodland

head very like that of modern birds

long, slender neck

claw on digit may have been used for opening termite nests

long, slim legs

feet with three forward-facing, clawed toes

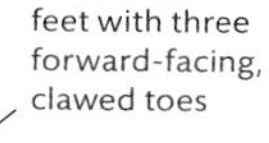

Length: 1m (2ft)	Weight: ~2.5kg (5½lb)	Diet: Insects, small reptiles

Group: COELUROSAURIA	Subgroup: Therizinosauridae	Time: 72–68mya

Therizinosaurus

THER-IZ-IN-OH-SAW-RUS

One of the strangest dinosaurs, the reconstruction of *Therizinosaurus* ("scythe lizard") is based on finds of other members of the family. Little is known of it apart from its incredibly long arms. These had three fingers that ended in curved, flat-sided claws, the first of which was longer than a man's arm. The claws seem too blunt for use in attack. Suggestions for their function have included that they were used in courtship or for raking plants. *Therizinosaurus* seems to have been an independent group, unrelated to Oviraptors as it had been suggested before.

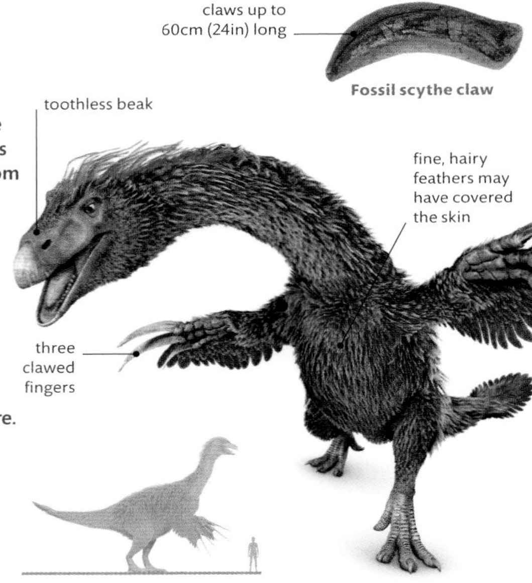

Fossil scythe claw

DESCRIBED BY
Maleev 1954
HABITAT
Woodland

Length: 11m (36ft)	Weight: Not calculated	Diet: Meat, or possibly plants

Group: COELUROSAURIA	Subgroup: Oviraptorosauria	Time: 76–68mya

Citipati

CHIT-I-PUH-TIH

Citipati is one of the oviraptorid dinosaurs, which got their name ("egg thief") because the first specimen was discovered on a nest of eggs, thought to be those of *Protoceratops*. However, dozens of new specimens (see right) show that these were actually oviraptorid eggs, and that oviraptorids brooded their nests just like chickens. Like most oviraptorids, *Citipati* had an unusual, short head, with a tall crest and a beak, which would have allowed them to have an omnivorous diet of plants and small prey. Fossils of its close relatives show that *Citipati* and other oviraptorosaurs were feathered.

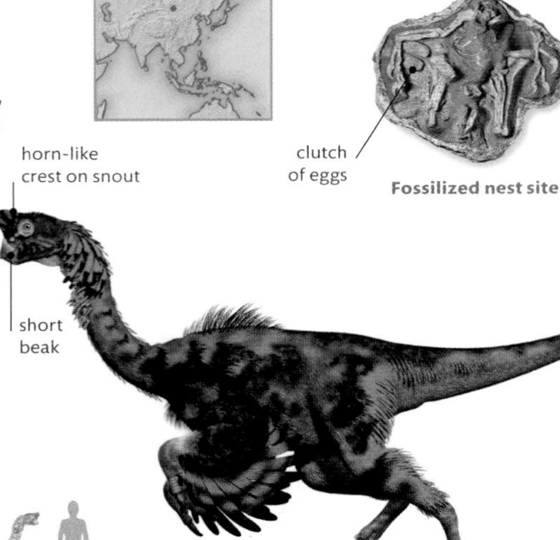

Fossilized nest site

DESCRIBED BY
Clark et al. 2001
HABITAT Semi-desert

Length: 2.5m (8ft)	Weight: ~105kg (231lb)	Diet: Plants and small prey

Group: COELUROSAURIA	Subgroup: Oviraptorosauria	Time: 130–122mya

Caudipteryx

CAWD-EE-TER-IKS

The discovery of this small, feathered dinosaur provided firm evidence of a theropod ancestry for birds. *Caudipteryx* ("tail feather") was a theropod with feathers covering its arms, most of its body, and its short tail. The feathers varied in structure - some were downy, and others were structured quills with shafts and veins. The feathers were symmetrical, showing that this animal did not fly. It is now thought that complex feathers of this kind evolved in an earlier, flightless theropod, and were inherited by both *Caudipteryx* and the ancestor of modern birds.

DESCRIBED BY Ji et al. 1998
HABITAT Lakesides

pointed beak with cluster of teeth at front of upper jaw

small, short skull with no teeth in lower jaw

short arms with symmetrical feathers

long, slim legs best-suited to running

bird-like feet with three forward-facing clawed toes

Length: 90cm (35in)	Weight: ~10kg (22lb)	Diet: Plants

Group: PARAVES	Subgroup: Dromaeosauridae	Time: 120mya

Microraptor

MIKE-ROW-RAP-TOR

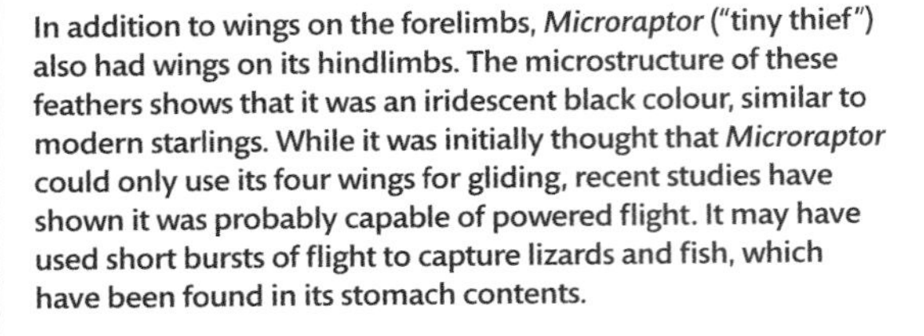

In addition to wings on the forelimbs, *Microraptor* ("tiny thief") also had wings on its hindlimbs. The microstructure of these feathers shows that it was an iridescent black colour, similar to modern starlings. While it was initially thought that *Microraptor* could only use its four wings for gliding, recent studies have shown it was probably capable of powered flight. It may have used short bursts of flight to capture lizards and fish, which have been found in its stomach contents.

DESCRIBED BY Xu et al. 2000
HABITAT Forests and lakes

wings with complex, branching feathers

wings of feathers on the legs as well as the arms

iridescent black plumage

diamond-shaped feather fan for stability

Length: 1m (3ft)	Weight: ~350g (12¼oz)	Diet: Lizards and fish

Group: PARAVES	Subgroup: Dromaeosauridae	Time: 108–98mya

Deinonychus

DIE-NO-NIKE-US

One of the most fearsome predators of the Cretaceous, *Deinonychus* ("terrible claw") was named for the sickle-shaped claws on the second toe of each foot. New research shows that these were most effective as puncture weapons. *Deinonychus* may have leapt onto and restrained prey with its sharp claws while it began to eat the animal alive. Although previously thought to be a pack hunter, based on a group found alongside a large *Tenontosaurus* (see p.209), it is now believed that *Deinonychus* died in a predator trap, attracted by the carcass of *Tenontosaurus*.

DESCRIBED BY Ostrom 1969
HABITAT Forests

Length: 3–4m (9¾–13ft)	Weight: ~70kg (155lb)	Diet: Herbivorous dinosaurs

Group: PARAVES	Subgroup: Dromaeosauridae	Time: 77–76mya

Dromaeosaurus

DROM-AY-OH-SAW-RUS

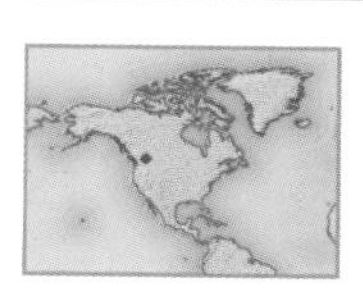

Dromaeosaurus ("running lizard") was the first sickle-clawed dinosaur to be discovered. However, the difficulty in reconstructing it from the few bones found meant that its true classification was only realized after *Deinonychus* was described. This dinosaur was smaller than *Deinonychus*, but otherwise very similar. Its body was slender, with long limbs and a large head. Its sharp claws would have been effective for restraining and piercing prey.

DESCRIBED BY Matthew and Brown 1922
HABITAT Forests, plains

Length: 1.8m (5½ft)	Weight: ~15kg (33lb)	Diet: Herbivorous dinosaurs

Group: PARAVES	Subgroup: Dromaeosauridae	Time: 83–76mya

Velociraptor

VEL-O-SEE-RAP-TOR

Many well-preserved skeletons of *Velociraptor* ("fast thief") have been found, making it the best-known member of its family. The main feature that sets it apart from other members of the family is its low, long, flat-snouted head. *Velociraptor* must have been a formidable predator: its jaws held about 80 very sharp teeth, and the second toe of each foot ended in a puncturing, sickle-shaped claw. The neck was S-shaped, and the hands bore three clawed fingers. It is probable that *Velociraptor* was a solitary predator. Traditionally reconstructed with scaly skin, recent analysis has shown that this dinosaur was covered in a downy coat or primitive feathers. It could not fly, but would have had a wing of feathers, much like modern birds.

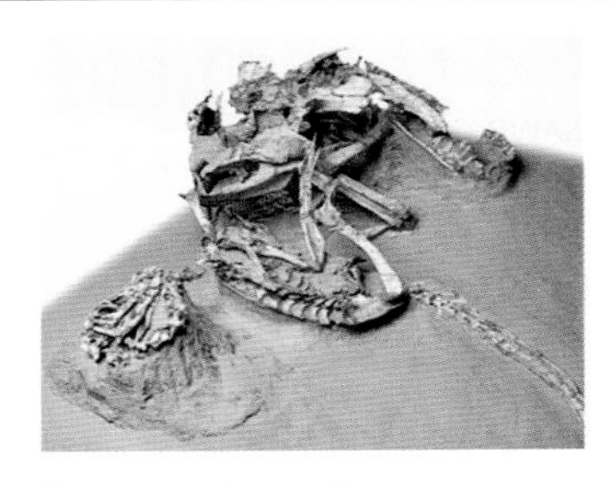

Preserved in death
One of the most famous fossil finds is a *Velociraptor* and *Protoceratops* locked together in battle at the time of their deaths.

DESCRIBED BY Osborn 1924
HABITAT Woodland

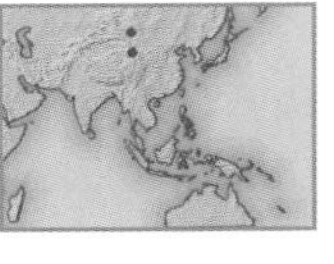

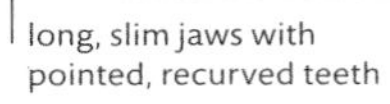

Fossil skull

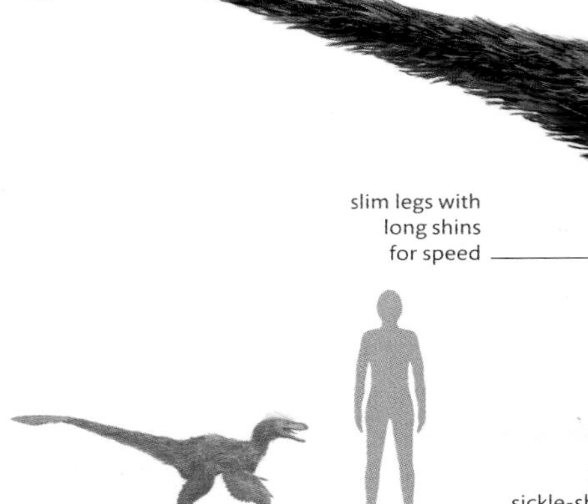

Length: 1.8m (6ft)	Weight: ~15kg (33lb)	Diet: Lizards, mammals, smaller dinosaurs

Group: PARAVES	Subgroup: Troodontidae	Time: 83–76mya

Saurornithoides

SAWR-OR-NITH-OID-EEZ

Only the skull, a few arm bones, and teeth of *Saurornithoides* ("lizard bird form") have been found to date, but its classification as a troodontid is now certain. Its long, narrow skull had a relatively large braincase. Its long, powerful arms ended in three-fingered hands capable of grasping prey.

relatively large braincase

long, narrow snout

jaw containing many sharp teeth

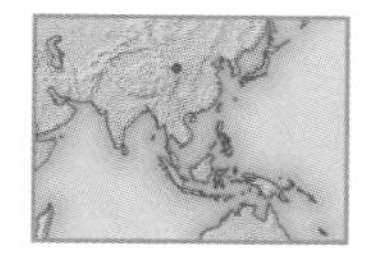

DESCRIBED BY Osborn 1924
HABITAT Plains

Length: 2m–3.5 (6½–11ft)	Weight: ~13–27kg (29–60lb)	Diet: Meat

Group: PARAVES	Subgroup: Troodontidae	Time: 83–76mya

Troodon

TROH-OH-DON

***Troodon* ("wounding tooth") was named for its sharp, serrated teeth. Its remains are very rare, and no complete skeleton has been found to date. Its reconstruction is based on the fossils found and details known from similar, closely related dinosaurs. It was probably an efficient hunter. It had long, slim hind legs, a large sickle-shaped claw on both second toes, and three long, clawed fingers that were capable of grasping prey. The sickle claws were smaller than in *Deinonychus* (see p.121) and *Velociraptor* (opposite). This has led some palaeontologists to suggest that they were mainly used for defence. *Troodon* could undoubtedly run quickly on its long legs, with its stiff tail held out behind for balance. Analysis of the slim skull shows that it had keen eyesight and may have had good hearing. Its braincase was also very large in proportion to its body size, indicating an unusually high level of intelligence for a dinosaur. *Troodon* eggs have been found in fossil nests.**

DESCRIBED BY Leidy 1856
HABITAT Plains

Length: 2m (6½ft)	Weight: ~50kg (110lb)	Diet: Meat, carrion

Group: PARAVES	Subgroup: Avialae	Time: 129–122mya

Confuciusornis

CON-FUSH-EE-US-OAR-NIS

Confuciusornis ("Confucius bird") was the first bird known to have a true, horny beak. It had a mixture of primitive features (clawed wing fingers and a flat breastbone) and modern adaptations, such as a deeper chest and shortened tail core. The foot claws were highly curved, and the large toe was reversed, indicating that it was a tree-dweller. Males had long tail feathers. Its beak turned up slightly at the end, leading to debate about its diet. *Confuciusornis* was most probably a side-shoot from the modern bird lineage.

three wing fingers with curved claws

toothless, horny beak with slight upward curve

DESCRIBED BY Hou et al. 1995
HABITAT Woodland

Length: 31cm (1ft)	Weight: Not calculated	Diet: Seeds, possibly fish

Group: AVIALAE	Subgroup: Ornithurae	Time: 84–78mya

Hesperornis

HESS-PER-OAR-NISS

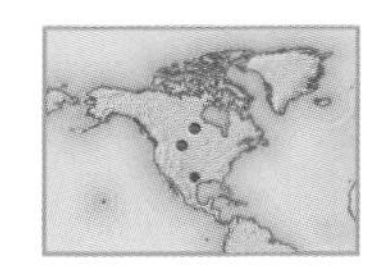

This large, toothed seabird lived a lifestyle very similar to that of modern penguins. It had lost the power of flight, and its vestigial wings were tiny and stubby. The head was long and low, with a long beak armed with sharp, pointed teeth for catching fish and other small marine creatures. *Hesperornis* ("western bird") was probably a powerful swimmer, using its large, webbed feet for propulsion, but may have moved clumsily on land. It probably nested on seashores.

legs set far back

DESCRIBED BY Marsh 1872
HABITAT Seashores

Length: 2m (6½ft)	Weight: Not calculated	Diet: Fish, squid

Group: AVIALAE	Subgroup: Ornithurae	Time: 95–83mya

Ichthyornis

IK-THEE-OAR-NISS

When the jaw of this toothed, primitive bird was re-examined in 1952, it was thought for a time, incorrectly, to belong to a juvenile mosasaur. The jaws and teeth of *Ichthyornis* ("fish bird") are indeed similar to those of a marine reptile. *Ichthyornis* was a seabird, similar in size and build to a modern seagull, but its head and beak were much larger. It had a well-developed, keeled breastbone, and a deep chest, suggesting it may have been a strong flyer. Its webbed feet had claws.

proportionately large head

long, horny bill filled with sharp teeth

webbed feet with short claws

DESCRIBED BY Marsh 1872
HABITAT Seashores

Length: 20cm (8in)	Weight: Not calculated	Diet: Fish

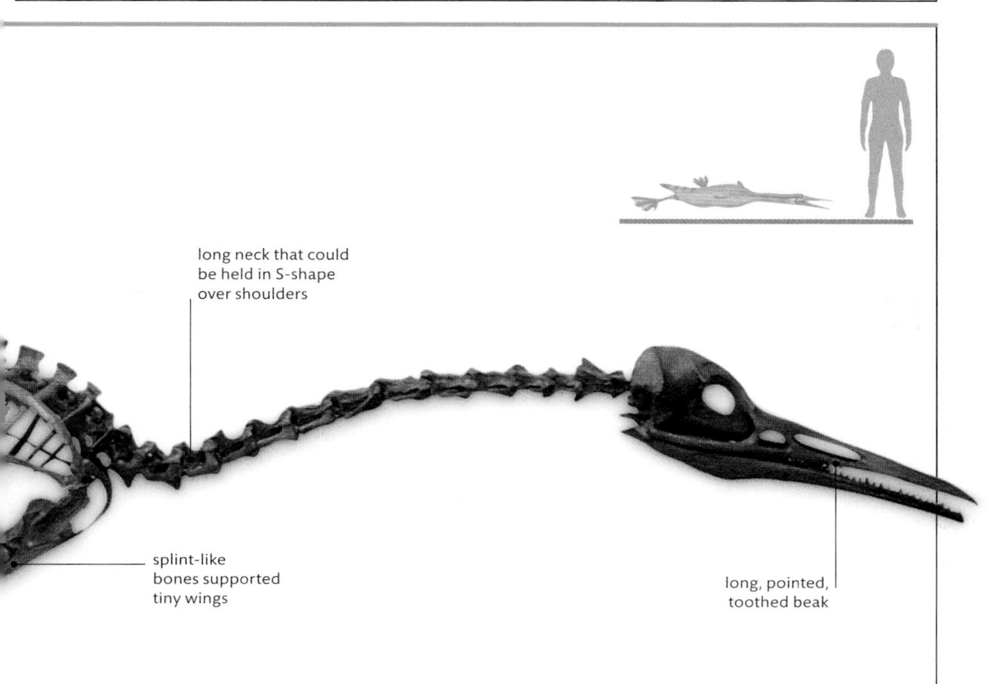

SAUROPODS

THE CRETACEOUS was the time of giants. It witnessed the rise of the appropriately named titanosaurs, which include the largest animals to ever walk on land. There is debate about which of the titanosaurs was the biggest, or even which measure is most important, but it is clear that *Argentinosaurus* from the Late Cretaceous was among them.

Titanosaurs became globally widespread, and have been found on every continent including Antarctica, but they were most common in the southern hemisphere. Communal nesting grounds show that they probably moved together in herds, and analyses of the internal structure of their bones suggest that they grew at exceptionally high rates to achieve their large sizes. Titanosaurs persisted until the end of the Cretaceous, when they went extinct with the rest of the non-avian dinosaurs.

In contrast, diplodocoids and brachiosaurids only survived partway throughout the Cretaceous, and neither group made it to the end-Cretaceous extinction. Even though both these groups were globally distributed in the Jurassic, their Early Cretaceous records are greatly restricted. Early Cretaceous brachiosaurids are known only from the southern part of North America and northwestern Gondwana, while diplodocoids were restricted to South America, Africa, and Europe. It is unclear why these groups declined and went extinct, but it was part of a broader restructuring of ecosystems across the Jurassic-Cretaceous boundary.

Group: SAUROPODA	Subgroup: Titanosauria	Time: 70–68mya

Saltasaurus

SALT-AH-SAW-RUS

Saltasaurus was named for the province in Argentina where it was first found, and was the first sauropod for which evidence of armour was found. The fossil remains consisted of a group of partial skeletons surrounded by thousands of bony plates. Most of these plates were tiny, but others were larger and may have ended in bony spikes. It is likely that these bony plates covered the back and sides of the dinosaur to form a protective covering. *Saltasaurus* had a long neck and a long, flexible tail. Its head was small, with a high skull. Like most sauropods, it would probably not have been able to rear up on its hind legs.

DESCRIBED BY Bonaparte and Powell 1980
HABITAT Woodland

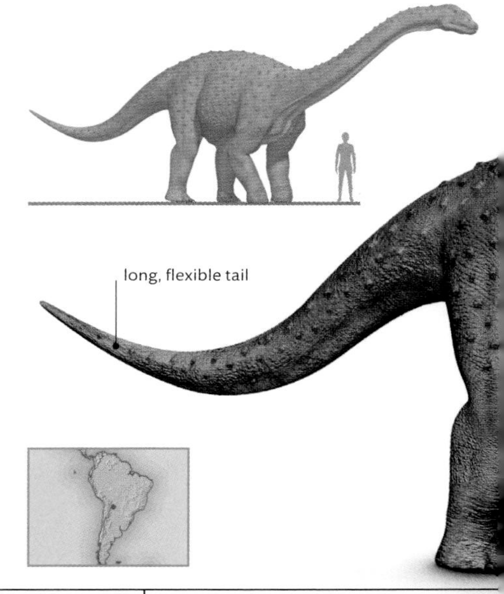

Length: 12m (39ft)	Weight: ~20 tonnes (19⅝ tons)	Diet: Plants

Group: SAUROPODA	Subgroup: Titanosauria	Time: 93–89mya

Argentinosaurus

AR-GEN-TEEN-OH-SAW-RUS

Only a few bones of *Argentinosaurus* (Argentina lizard) have been found to date, including some enormous vertebrae from the back, which were over 1.5m (5ft) wide in cross-section. Other bones found were the sacrum, a fibula, and a few ribs. Because of the sparsity of fossil evidence, very little is known about *Argentinosaurus*. However, it appears to have been the heaviest dinosaur ever known. As in other titanosaurs, the neck and tail would have been very long and thin, and the skull is likely to have been small and triangular.

DESCRIBED BY Bonaparte and Coria 1993
HABITAT Forested areas

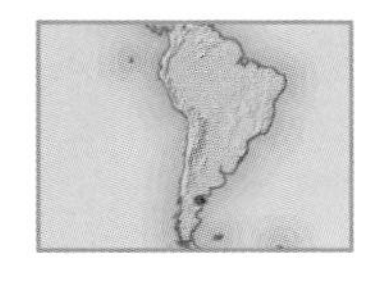

long, slim neck

body thought to resemble that of other titanosaurs

long tail

elephant-like, thick limbs with clawed toes

Length: 30m (98ft)	Weight: ~96 tonnes (106 tons)	Diet: Conifers

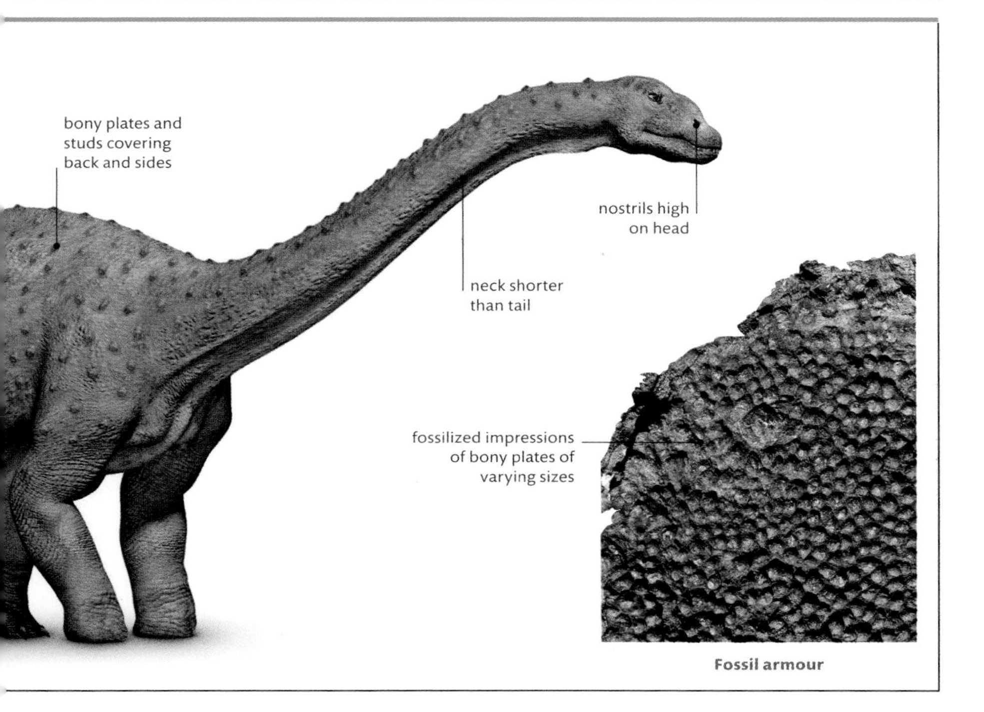

Fossil armour

ORNITHISCHIANS

ALL MAJOR lineages of ornithischians survived into the Cretaceous and continued to thrive until the end-Cretaceous extinction.

While thyreophorans were larger and more successful in the Jurassic, cerapodans (the group including marginocephalians and ornithopods) took over in the Cretaceous. Ankylosaurs survived until the end of the period, but stegosaurs went extinct in the Early Cretaceous.

Iguanodontians, with distinctive thumb spikes, evolved from small bipedal ornithopods in the Early Cretaceous. In the Late Cretaceous, they gave rise to hadrosaurs that had thousands of highly specialized teeth for grinding vegetation. Hadrosaurs were incredibly successful and spread out across Laurasia, dispersing across oceans to South America and Africa towards the end of the Cretaceous.

Both marginocephalian groups - the pachycephalosaurs and ceratopsians - diversified throughout the Cretaceous. Pachycephalosaurs developed elaborate ornamented domes and spread throughout Asia and North America. Ceratopsians remained small until the Late Cretaceous, when they increased in size, exploded in diversity, and formed vast herds that roamed across North America. Some of the big Late Cretaceous ceratopsians, such as *Torosaurus*, had the largest skulls of any land animal. Ceratopsians also developed a complex battery of teeth, but they employed a slicing method of chewing. Two groups of ceratopsians evolved: the ornately-frilled centrosaurines, and the long-horned chasmosaurines. Neither of these groups would ever reach Asia, but the reason for this remains unclear.

Group: ANKYLOSAURIA	Subgroup: Ankylosauridae	Time: 68–66mya

Ankylosaurus

AN-KIE-LOH-SAW-RUS

Ankylosaurus ("fused lizard") has aptly been described as a living tank. Its stocky body, neck, and head were protected by thick bands of armour-plating. The skin was thick and leathery, and was studded with hundreds of oval bony plates and rows of spikes. A pair of long spikes stuck out from the back of the head, and the cheekbones were drawn out into another pair of spikes on the face. The tail was armed with a bony club, which could be swung with great force. The legs were strong, although short, and the body was wide and squat. The face was also broad, with a blunt snout that ended in a toothless beak.

DESCRIBED BY Brown 1908
HABITAT Woodland

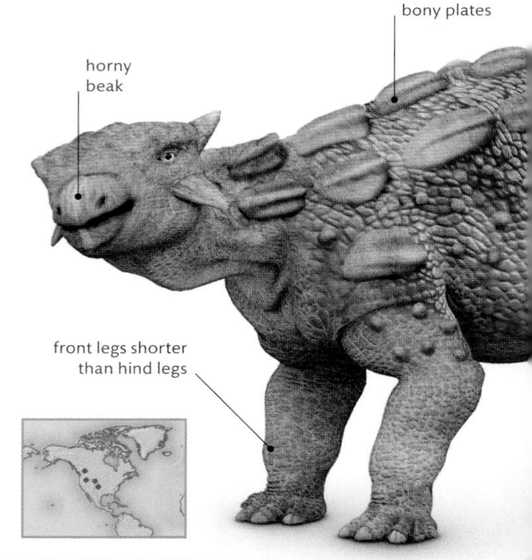

Length: 7.5–10.5m (25–35ft)	Weight: ~4.5–7 tonnes (4⅜–6⅞ tons)	Diet: Plants

Group: THYREOPHORA	Subgroup: Ankylosauria	Time: 105–99mya

Kunbarrasaurus

KOON-BAR-AH-SAW-RUS

This small, heavily-built and thickly armoured ankylosaur is named for the Mayi (language of the Wunumara indigenous people of Queensland, Australia) word for "shield". It had rows of small, bony plates running down its back and triangular spikes over the hips. There were also large bony plates over the neck and shoulders. The head was box-shaped, with a very narrow snout ending in a horny beak. Four small horns jutted out from the back of the face.

DESCRIBED BY Leahey et al. 2015
HABITAT Scrubby and wooded plains

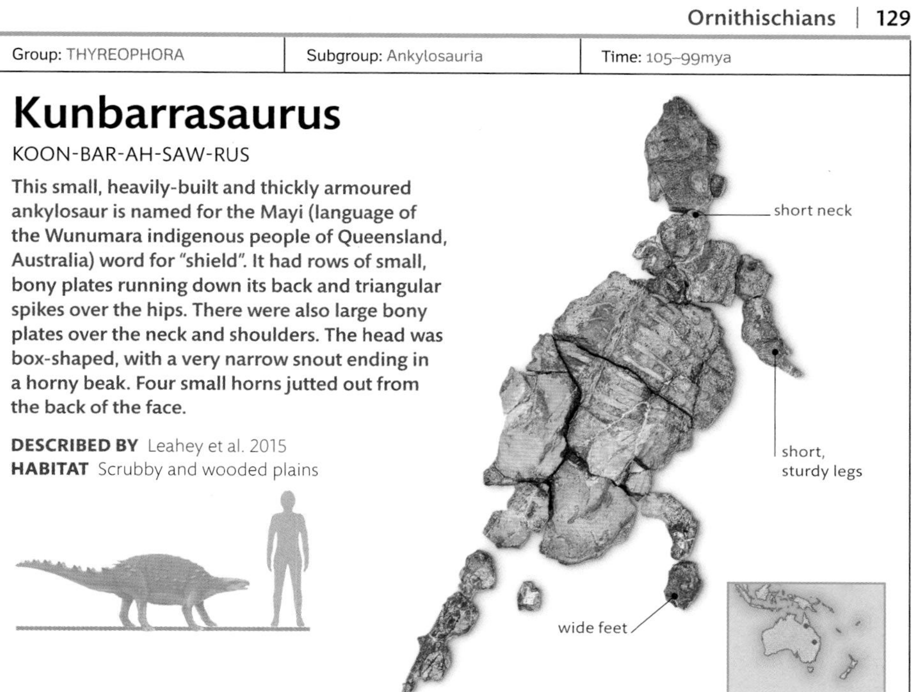

Length: 3m (9¾ft)	Weight: Not calculated	Diet: Low-lying vegetation

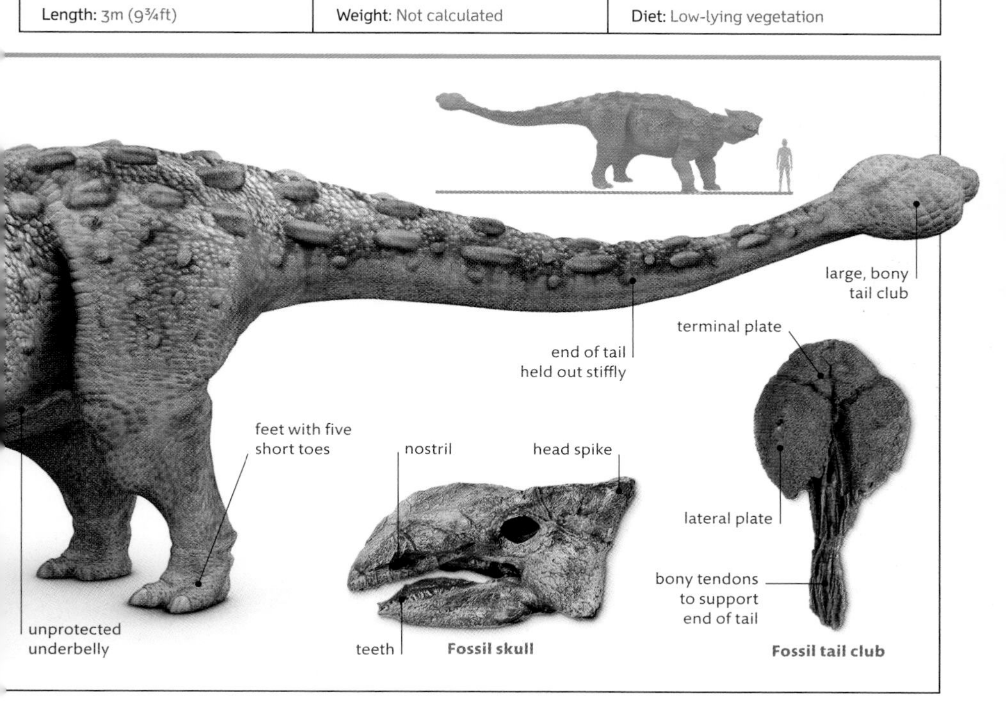

Fossil skull

Fossil tail club

Group: ANKYLOSAURIA	Subgroup: Ankylosauridae	Time: 144–112mya

Gastonia

GAS-TONE-EE-A

Gastonia (named after palaeontologist and cast-maker Robert Gaston) had a mixture of ankylosaur and nodosaur features. This herbivore's defensive armour was impressive: there were four horns on the head, and bony rings covering the neck; rows of spikes covered the back and the flanks, and fused, bony armoured plates protected the hips. The tail, which could lash from side to side, had rows of triangular blades along each side. Running down both sides of the spine, long spikes curved up and outwards, forming a formidable defensive shield against attack by predators.

curved dorsal spines

horny beak

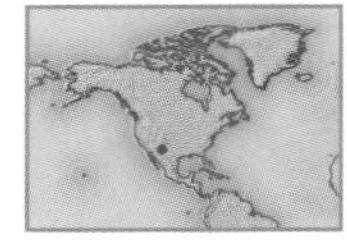

DESCRIBED BY Kirkland 1998
HABITAT Woodland

Length: 5m (16ft)	Weight: ~1 tonne (1 ton)	Diet: Plants

Group: ANKYLOSAURIA	Subgroup: Ankylosauridae	Time: 77–76mya

Euoplocephalus

YOU-OH-PLO-SEF-AH-LUSS

Built like a military tank, the back of *Euoplocephalus* ("well-armoured head") was embedded with bands of armour and bony studs. Fused plates covered the neck, and triangular horns protected the shoulders, base of tail, and the face. A large ball of fused bone at the end of the tail acted as a club that could be swung at attacking predators.

DESCRIBED BY Lambe 1910
HABITAT Woodland

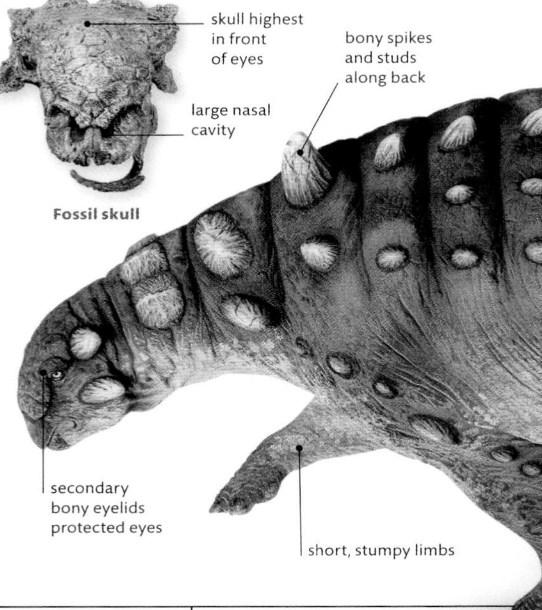

Fossil skull

Length: 6m (20ft)	Weight: ~2 tonnes (2 tons)	Diet: Plants

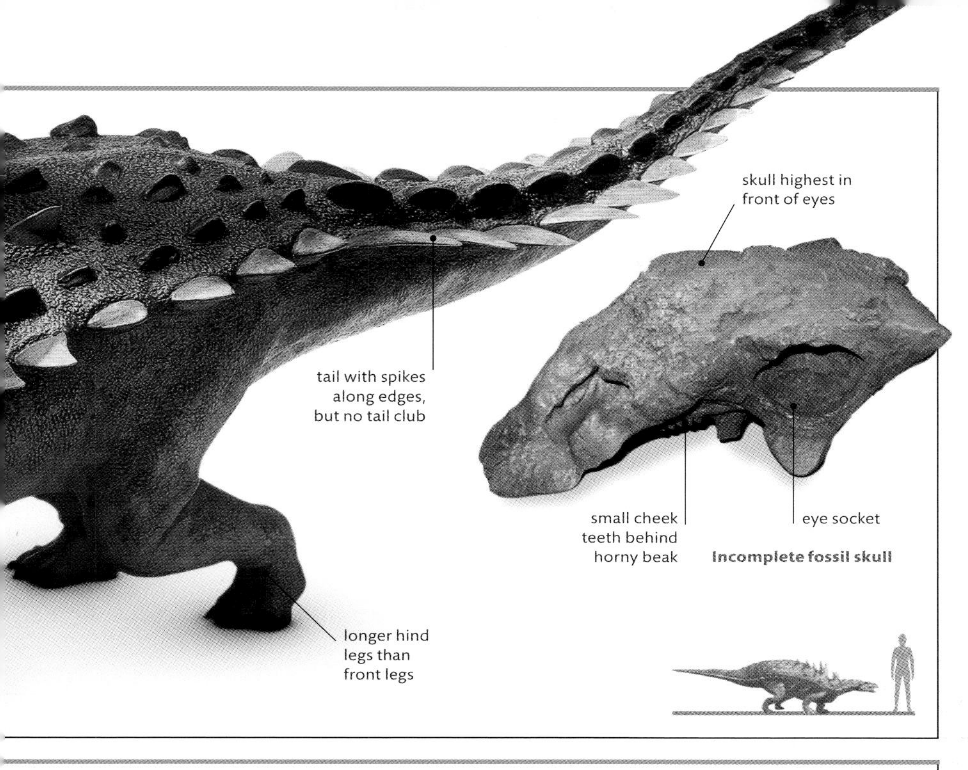
skull highest in
front of eyes
tail with spikes
along edges,
but no tail club
small cheek
teeth behind
horny beak
eye socket
Incomplete fossil skull
longer hind
legs than
front legs

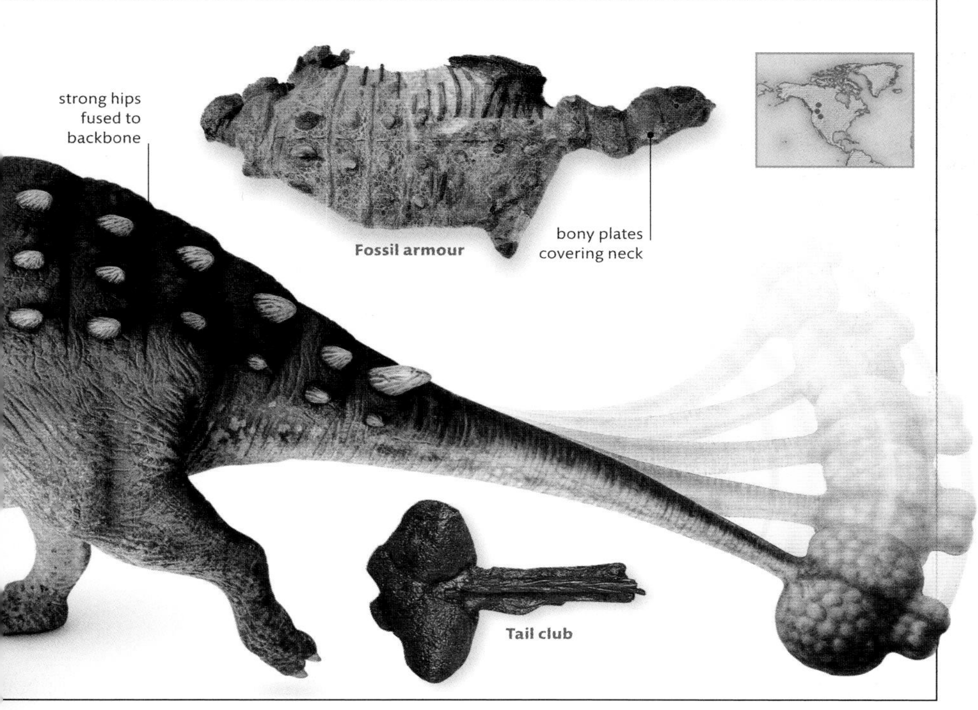
strong hips
fused to
backbone
bony plates
covering neck
Fossil armour
Tail club

Group: ANKYLOSAURIA	Subgroup: Nodosauridae	Time: 73–71mya

Edmontonia

ED-MON-TONE-EE-AH

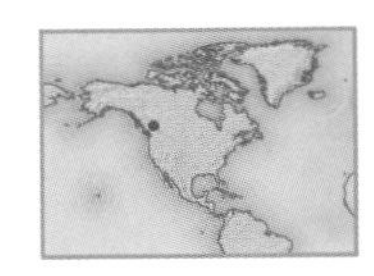

This ankylosaur (whose name means "from Edmonton") had a bulky body, four short, thick legs, wide feet, and a short neck. The back and tail were covered with rows of bony plates (scutes) and spikes. The shoulders of this dinosaur were particularly well-armoured with long spikes and large scutes. Fossil scutes that have been found are not symmetrical, meaning that they would not have projected straight out from the hide, but at an angle. *Edmontonia*'s skull was long and flat, with large nasal cavities and weak jaws. The front of the snout was formed into a horny, toothless beak, with small, weak cheek teeth inside the mouth. Some palaeontologists have suggested that *Edmontonia* is more likely to have used its shoulder spikes for defence or sexual display rather than combat.

DESCRIBED BY Sternberg 1928
HABITAT Woodland

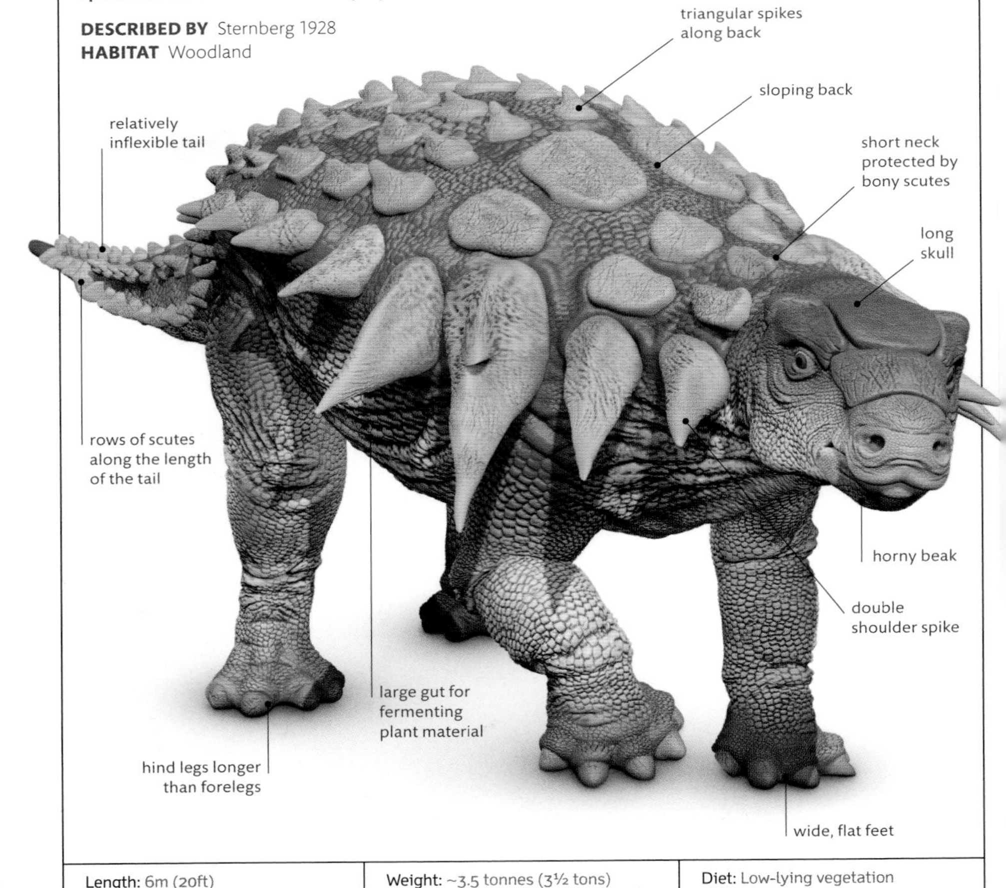

Length: 6m (20ft)	Weight: ~3.5 tonnes (3½ tons)	Diet: Low-lying vegetation

Group: MARGINOCEPHALIA	Subgroup: Pachycephalosauria	Time: 77–76mya

Stegoceras

STEG-OH-SER-AS

A fast runner, *Stegoceras* (meaning "roof horn") had a body designed to withstand vigorous head-butting activity. When charging, its head was lowered at right angles, balanced by its neck, body, and tail held in a straight line. The skullcap was thickened into a dome of solid bone, and the grain of bone was angled to the surface, enabling it to withstand impact more effectively. Two types of skull have been found – ones with low, flat domes, and others with tall, rounded ones. These variations probably reflect differences between juveniles and adults. *Stegoceras* had small, serrated teeth that were ideal for shredding plant material.

DESCRIBED BY Lambe 1902
HABITAT Upland forests

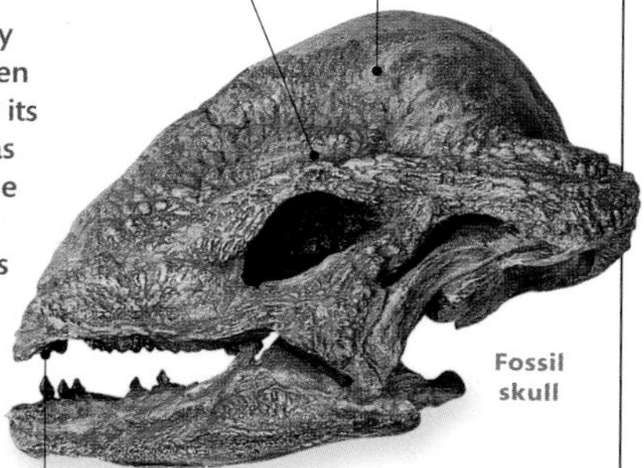

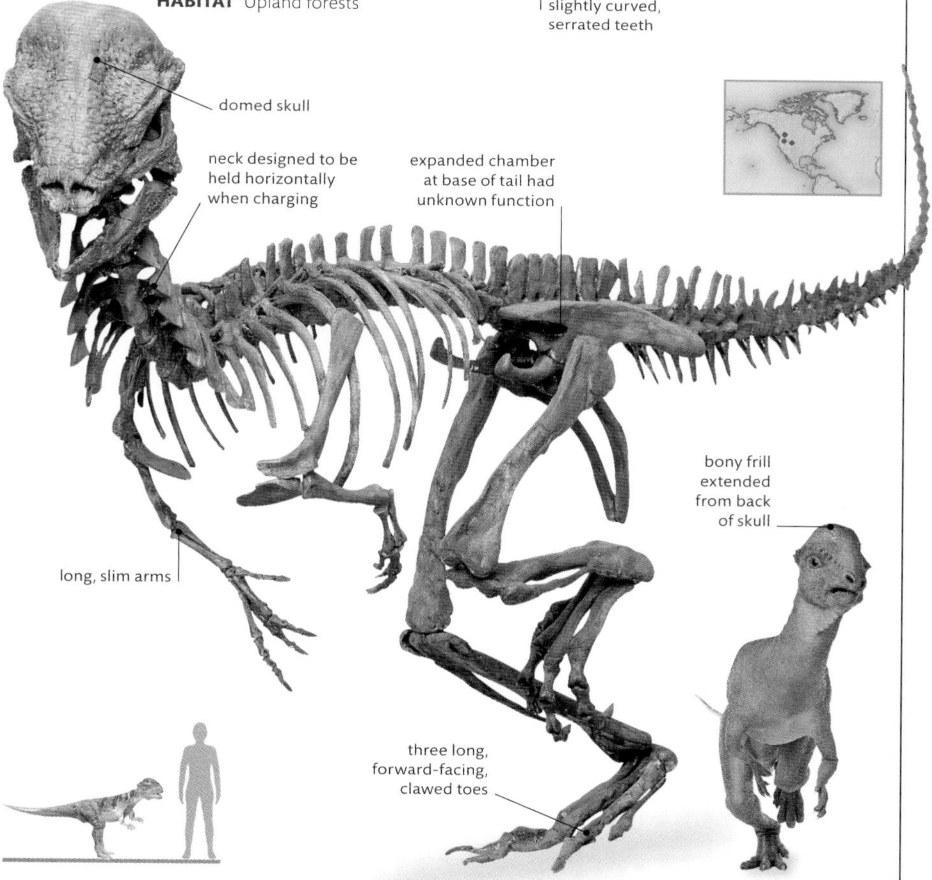

Length: 2m (6½ft)	Weight: ~54kg (120lb)	Diet: Leaves, fruit

Group: MARGINOCEPHALIA	Subgroup: Pachycephalosauria	Time: 70–66mya

Pachycephalosaurus

PACK-EE-SEF-AL-OH-SAW-RUS

It is easy to see why *Pachycephalosaurus* ("thick-headed lizard") was so named. The top of the dinosaur's skull was thickened into a dome of solid bone 25cm (10in) thick. It is thought that *Pachycephalosaurus* used its head as a battering ram for defence or in territorial fighting. It had a beak, and clusters of bony knobs surrounded the snout. Further groups of round knobs encircled the back of the head. Its relatively slim leg and foot bones indicate that this dinosaur may have been able to run more quickly than its heavy build might suggest. *Pachycephalosaurus* is the largest member of the Pachycephalosauridae family of dinosaurs discovered so far, and was the last to exist before the extinction of the dinosaurs at the end of the Cretaceous Period.

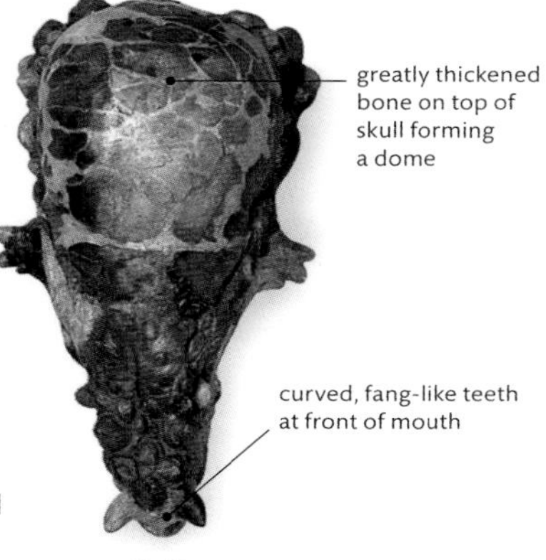

Skull

DESCRIBED BY Brown and Schlaikjer 1943
HABITAT Forests

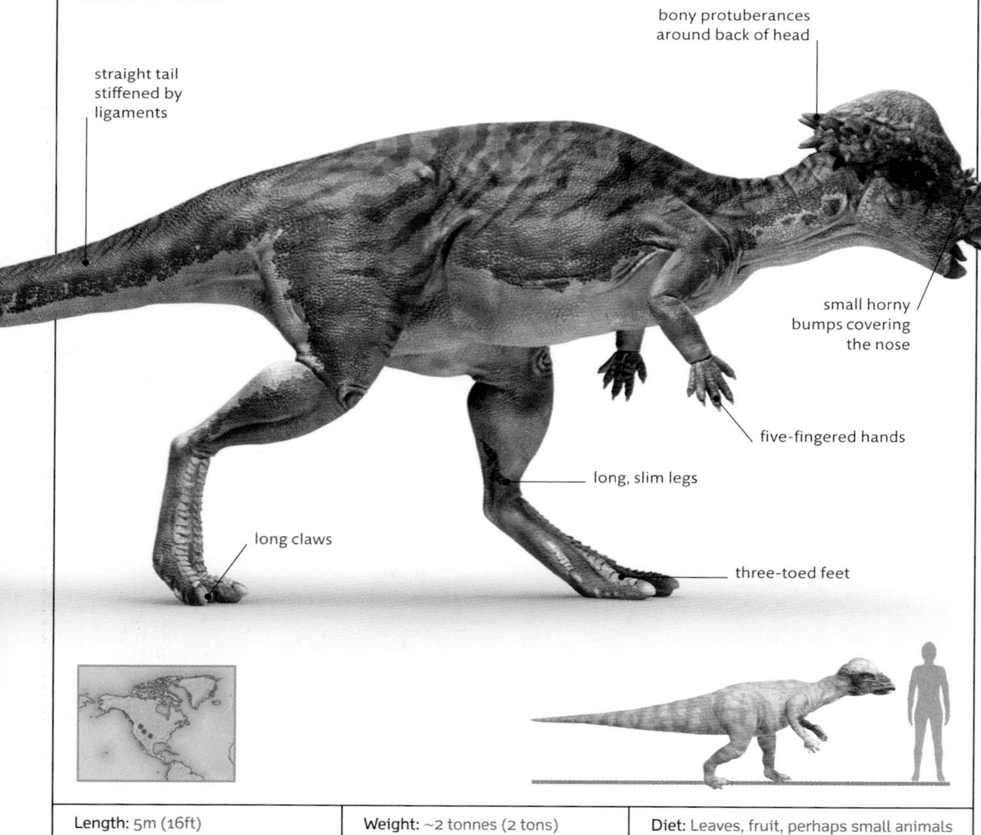

Length: 5m (16ft)	Weight: ~2 tonnes (2 tons)	Diet: Leaves, fruit, perhaps small animals

Group: MARGINOCEPHALIA	Subgroup: Ceratopsia	Time: 126–101mya

Psittacosaurus

SIT-A-KOH-SAW-RUS

Its characteristic square skull and curved beak were the features that led to this dinosaur being given its name, which means "parrot lizard". Its cheek bones formed a pair of horns, believed to have been used for fighting or sexual display. Its hind legs were long and thin, suggesting that it was a fast bipedal runner. The long toes with blunt claws were perhaps used for digging. Its long tail was stiffened by bony tendons along its length.

teeth behind toothless beak

Fossil skull

horn

DESCRIBED BY Osborn 1923
HABITAT Desert and scrubland

tail to maintain balance

four clawed toes on hind feet

Length: 2m (6¼ft)	Weight: ~80kg (175lb)	Diet: Plants

Group: MARGINOCEPHALIA	Subgroup: Ceratopsia	Time: 83–76mya

Protoceratops

PRO-TOE-SER-A-TOPS

Protoceratops ("before the horned faces") is very well-known from the many specimens discovered. It had a broad neck frill at the back of its skull. There was a small nasal horn between the eyes and numerous teeth in the upper jaw, similar to other ceratopsians. *Protoceratops* had slender legs and it was unlikely to have been a fast runner.

head frill, which grew with age

Adult skull

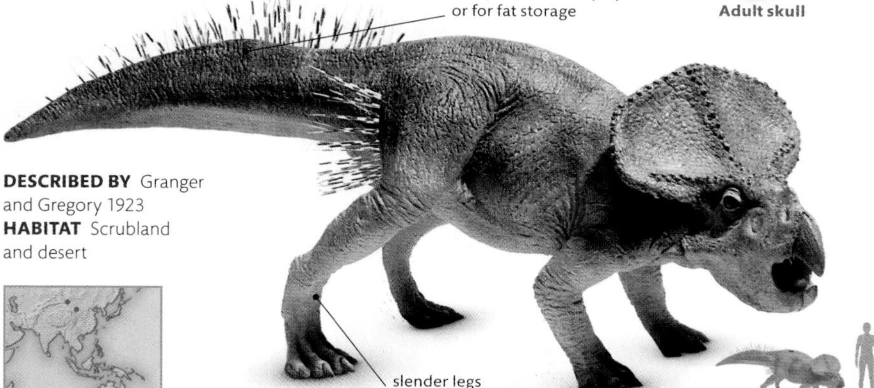

DESCRIBED BY Granger and Gregory 1923
HABITAT Scrubland and desert

Length: 1.8m (6ft)	Weight: ~180kg (400lb)	Diet: Plants

Group: CERATOPSIA	Subgroup: Centrosaurinae	Time: 77–76mya

Centrosaurus

SEN-TROH-SAW-RUS

The most distinctive feature of *Centrosaurus* ("pointed lizard") was the long horn on the snout. There were also two small brow horns and a neck frill standing up behind the head. This neck frill had a wavy edge and spines. *Centrosaurus* had a massive body, a short tail, and sturdy legs. Many *Centrosaurus* and *Styracosaurus* have been found together in a single bone bed, having died during a drought around a watering hole.

DESCRIBED BY Lambe 1904
HABITAT Woodland

wavy edge
brow horn for display and defence
skin-covered holes to reduce weight of neck frill
saw-edged teeth
horny beak

Length: 6m (20ft)	Weight: ~3 tonnes (3 tons)	Diet: Low-lying plants

Group: CERATOPSIA	Subgroup: Centrosaurinae	Time: 77–76mya

Styracosaurus

STY-RAK-OH-SAW-RUS

One of the most spectacular of the horned lizards, *Styracosaurus* ("spiked lizard") had six long spikes on the back edge of its neck frill, and smaller spikes around them. There was a large horn on the snout that pointed upwards and forwards. Its snout was very deep, and its nostrils seem to have been unusually large, although the reason for this is unknown. All four feet had five fingers or toes with claw-like hooves. Its teeth grew continuously to replace worn ones, and had a shearing action for slicing through tough plant material.

DESCRIBED BY Lambe 1913
HABITAT Open woodland

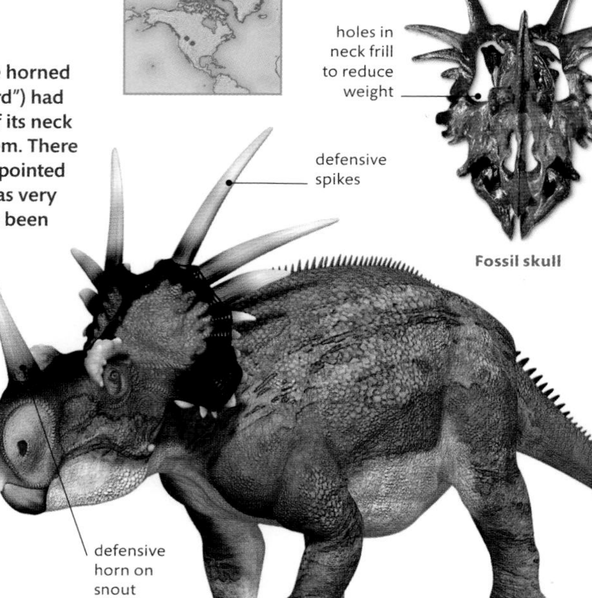

Fossil skull

Length: 5.2m (17ft)	Weight: ~2.8 tonnes (2¾ tons)	Diet: Ferns and cycads

Group: CERATOPSIA	Subgroup: Chasmosaurinae	Time: 77–76mya

Chasmosaurus

KAZ-MOH-SAW-RUS

Chasmosaurus ("opening lizard") was a typical frilled, horned dinosaur. It had a large body with four stocky legs. Its most distinctive feature – its enormous neck frill – was likely to have been brightly coloured in life for sexual display. The frill was so long that it reached over the shoulders. Its bony structure was lightened by two large holes that were covered with skin. Triangular, bony protrusions ran along the edge of the frill. *Chasmosaurus* had a small nose horn, and two blunt brow horns, which were of different lengths in different species or sexes. There was also a parrot-like beak at the front of the snout. The skin was covered in knob-like bumps with five or six sides.

DESCRIBED BY Lambe 1914
HABITAT Woodland

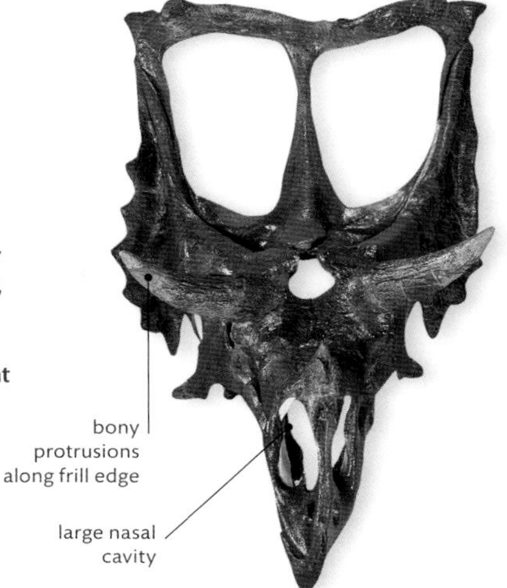

Front view of skull

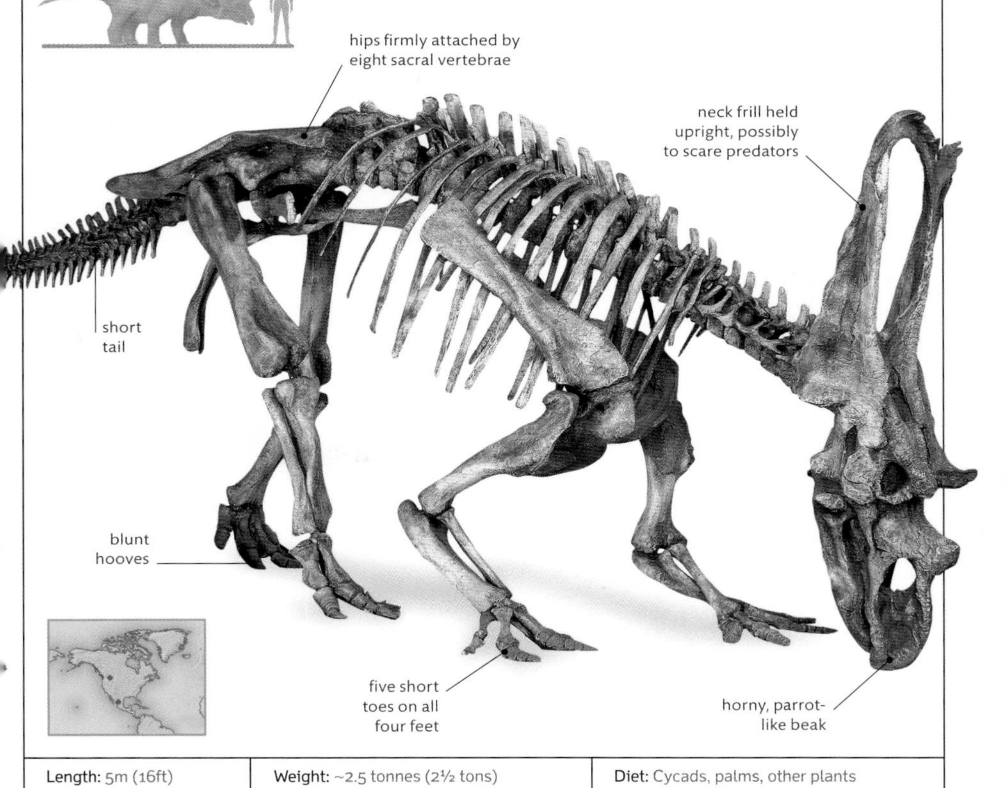

Length: 5m (16ft)	Weight: ~2.5 tonnes (2½ tons)	Diet: Cycads, palms, other plants

Group: CERATOPSIA	Subgroup: Chasmosaurinae	Time: 68–66mya

Triceratops

TRY-SER-A-TOPS

Probably the best-known of the horned dinosaurs, *Triceratops* (meaning three-horned face") lived throughout the lands that now form North America at the end of the Cretaceous Period. The nose horn was short and thick, and the two long brow horns, each over 1m (3¼ft) long, curved forwards and slightly outwards over the snout. There were pointed studs set around the edge of the neck frill for further protection and ornamentation. Many of the fossil skulls found have evidence of scarring. This suggests that *Triceratops* may have fought in territorial or mating battles by locking horns with rivals.

DESCRIBED BY Marsh 1889
HABITAT Woodland

TRICERATOPS SKULL

The solid structure of *Triceratops'* skull has meant that it survived fossilization better than most other dinosaur skulls. More than 50 *Triceratops* skulls have been found to date.

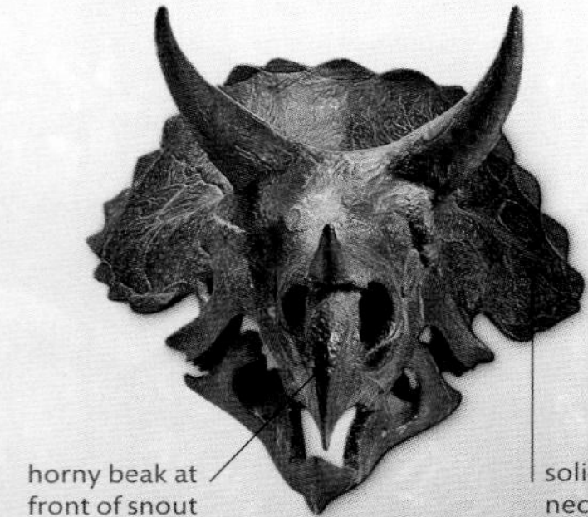

Length: 9m (30ft)	Weight: ~4.5–10 tonnes (4⅜–9⅞ tons)	Diet: Plants

Reconstruction of skeleton

Group: CERATOPSIA	Subgroup: Chasmosaurinae	Time: 75–73mya

Pentaceratops

PEN-TAH-SER-A-TOPS

Pentaceratops ("five-horn face") was so named because, in addition to the usual straight horn on the snout and two curved brow horns, its cheeks had spike-like protrusions. *Pentaceratops'* most unusual feature is the size of its head - a skull reconstructed in 1998 was over 2.3m (7.6ft) long. Its neck frill was also enormous, and edged with triangular, bony projections. The body was stockily built, with a short, pointed tail, a wide body, and hoof-like claws.

Skull fragment

DESCRIBED BY Osborn 1923
HABITAT Wooded plains

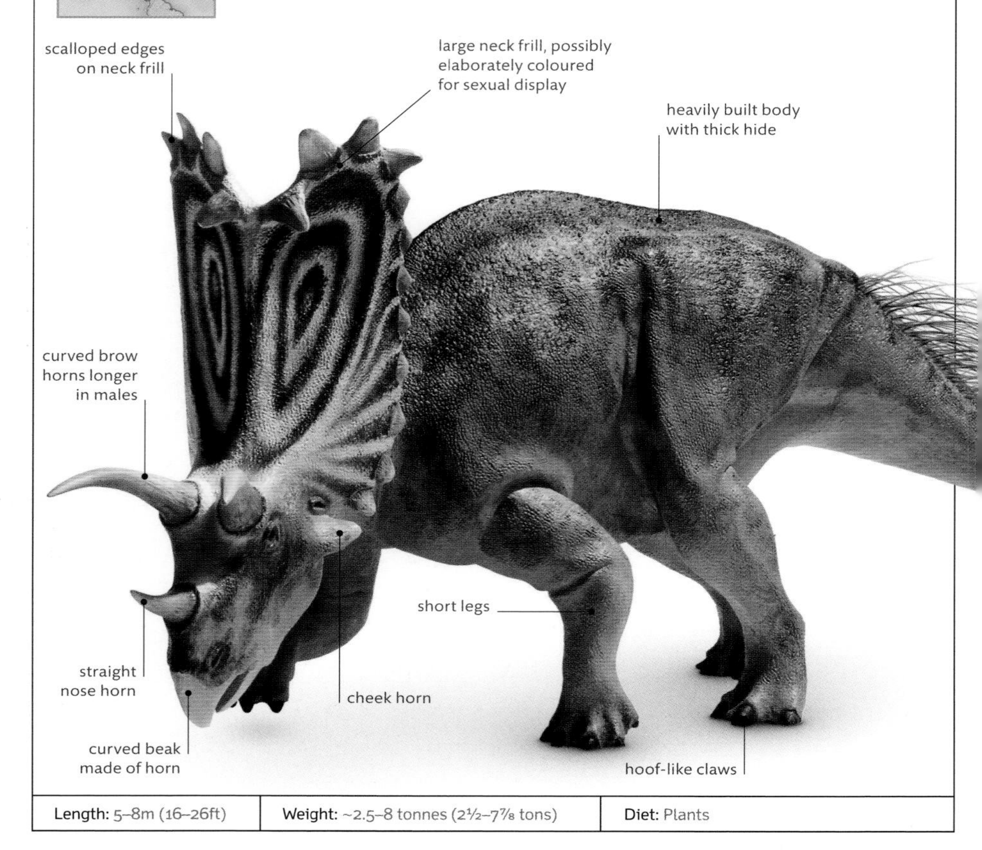

Length: 5–8m (16–26ft)	Weight: ~2.5–8 tonnes (2½–7⅞ tons)	Diet: Plants

Group: ORNITHISCHIA	Subgroup: Ornithopoda	Time: 132–129mya

Hypsilophodon

HIP-SILL-OH-FO-DON

This primitive ornithischian, whose name means "*Hypsilophus* tooth", is generally believed to have been a fully terrestrial dinosaur that was capable of running at high speeds. It walked on its hind legs; its short thigh bones and long shins meant that it was capable of taking long strides. The tail, stiffened by a network of bony ligaments, helped to balance the animal as it ran. Its small head had a horny beak and large eyes. The jaws held 28 or 30 cheek teeth that were self-sharpening, and the mouth had cheek pouches. Groups of fossils found together in bone beds suggest that this was a herding dinosaur.

DESCRIBED BY Huxley 1869
HABITAT Forests

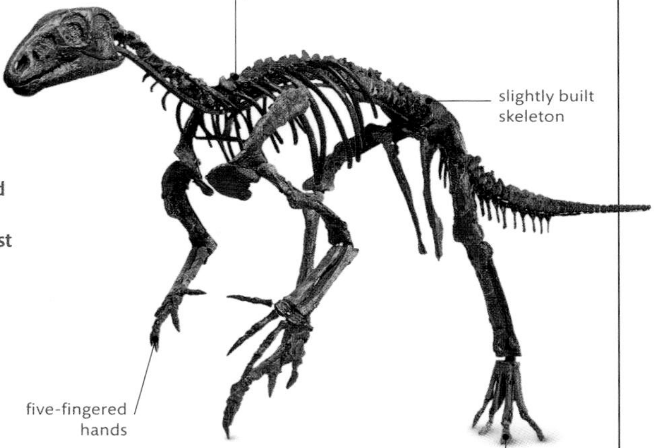

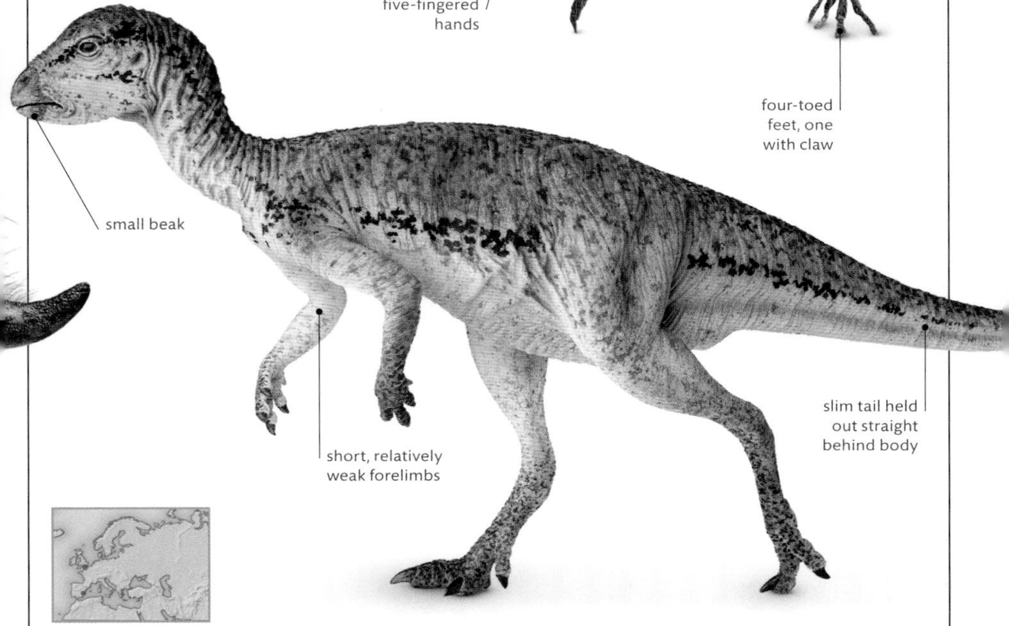

Length: 2m (6½ft)	Weight: ~68kg (150lb)	Diet: Plants

Group: ORNITHOPODA	Subgroup: Iguanodontia	Time: 132–129mya

Iguanodon

IG-WAH-NOH-DON

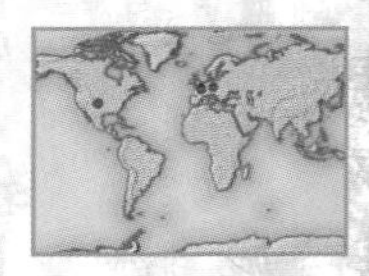

Iguanodon ("iguana tooth") was thought to have been capable of bipedal walking, but recent analysis suggests that it was strictly quadrupedal. Its hind legs were thick and column-like, while the front legs were considerably shorter and thinner. The middle three fingers on each hand were joined together by a pad of skin and the fifth finger could curl to grasp food. The thumb was armed with a long spike. *Iguanodon* could chew food; it had a hinged upper jaw that allowed the teeth in the upper jaw to grind over those in the lower jaw.

DESCRIBED BY Mantell 1825
HABITAT Woodland

Length: 9m (30ft)	Weight: ~4–5 tonnes (3⅞–4⅞ tons)	Diet: Plants

Fossil skull

Fossil thumb spike

Group: ORNITHOPODA	Subgroup: Iguanodontia	Time: 129–125mya

Ouranosaurus

OO-RAH-NOH-SAW-RUS

The most remarkable feature of *Ouranosaurus*, whose name means "brave lizard", was the row of spines growing out from the spinal and caudal vertebrae. They ran from the shoulders to halfway along the tail. Some palaeontologists think they supported a sail that was most likely used for sexual display. *Ouranosaurus* had a pair of bony bumps between the eyes, and a horny beak.

DESCRIBED BY Taquet 1972
HABITAT Tropical plains and forest

pair of bony bumps formed head crest

horny beak

thumb spike

Length: 7m (23ft)	Weight: ~4 tonnes (4 tons)	Diet: Leaves, fruit, and seeds

Group: HADROSAURIDAE	Subgroup: Saurolophinae	Time: 83–70mya

Maiasaura

MAY-A-SAW-RA

This dinosaur's name means "good mother lizard". It was so named because remains were found close to fossilized nests scooped out of mud. The nests were each about 2m (6½ft) across and contained eggs arranged in circular layers. Remains of juvenile dinosaurs at various stages of development showed that the parents brought food to their young at the nest site for a considerable time. It appears that *Maiasaura* nested each year in large herds. It had numerous cheek teeth adapted to grinding tough plant material.

DESCRIBED BY Horner and Makela 1979
HABITAT Coastal plains

long hind legs

stiff, narrow tail

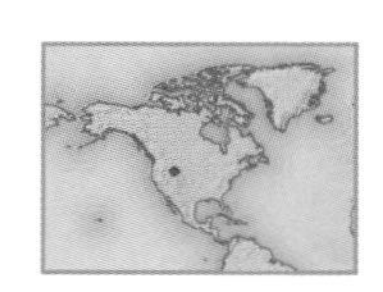

Skeleton of juvenile

Length: 9m (30ft)	Weight: ~5 tonnes (4⅞ tons)	Diet: Leaves

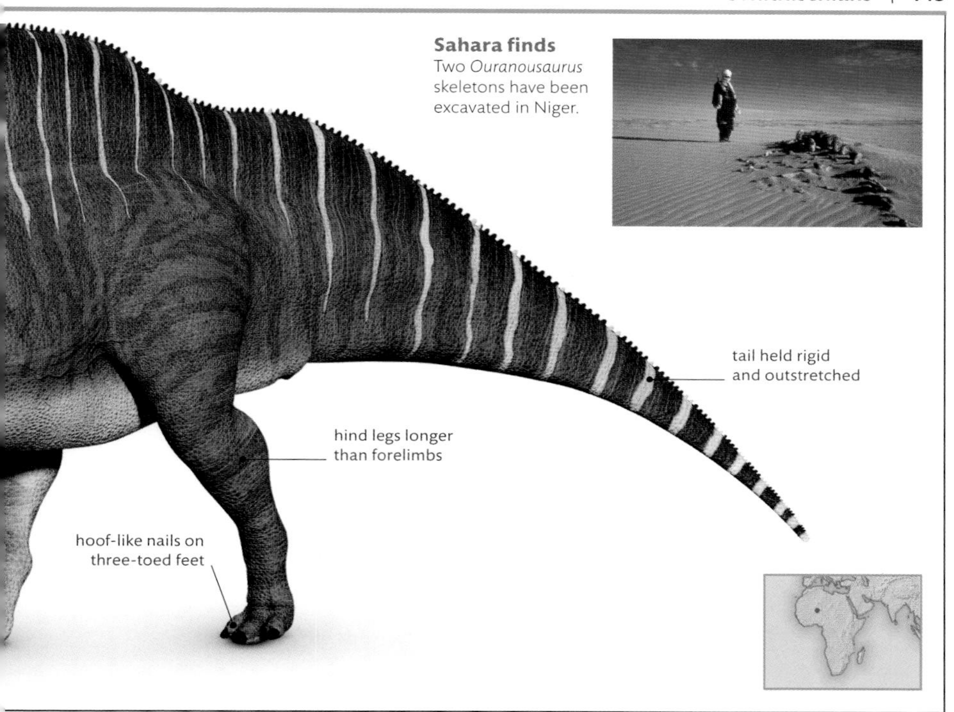

Sahara finds
Two *Ouranousaurus* skeletons have been excavated in Niger.

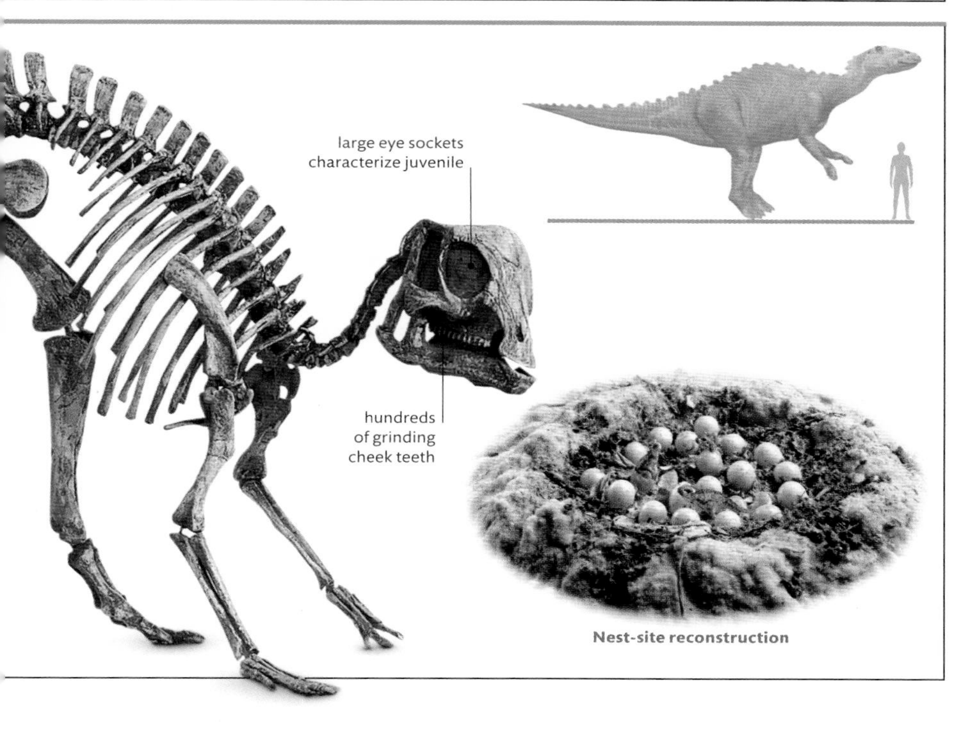

Nest-site reconstruction

Group: HADROSAURIDAE	Subgroup: Lambeosaurinae	Time: 77–76mya

Corythosaurus

KOH-RITH-OH-SAW-RUS

Corythosaurus **("Corinthian helmet lizard") was named for the distinctive hollow crest on top of its head. The crest was made up of greatly altered nasal bones, and the hollows were expanded nasal passages. Models of the crest have been made, which produce a booming foghorn-like sound when blown through, adding to the current theory that the crest served as a resonating device for signalling amongst herds.** ***Corythosaurus*** **was a low browser that spent most of its time on all fours. It had a broad toothless beak.**

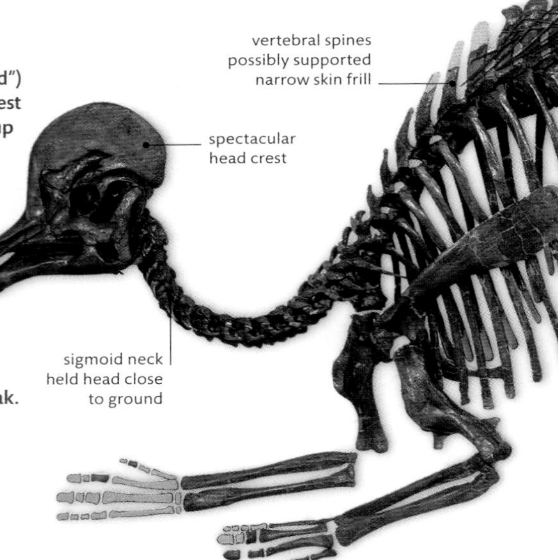

DESCRIBED BY Brown 1914
HABITAT Forests

Length: 10m (33ft)	Weight: ~4.5 tonnes (4⅜ tons)	Diet: Leaves, seeds, pine needles

Group: HADROSAURIDAE	Subgroup: Lambeosaurinae	Time: 77–73mya

Parasaurolophus

PA-RA-SAW-ROL-OFF-US

This dinosaur, whose name means "near ***Saurolophus*****", is instantly recognizable from its trombone-like head crest. This was up to 1.8m (6ft) long, and was probably used for display and as a resonating device for sound signalling. It was thought that a frill of skin connected the crest to the neck, but this is now considered unlikely.**

DESCRIBED BY Parks 1922
HABITAT Woodland

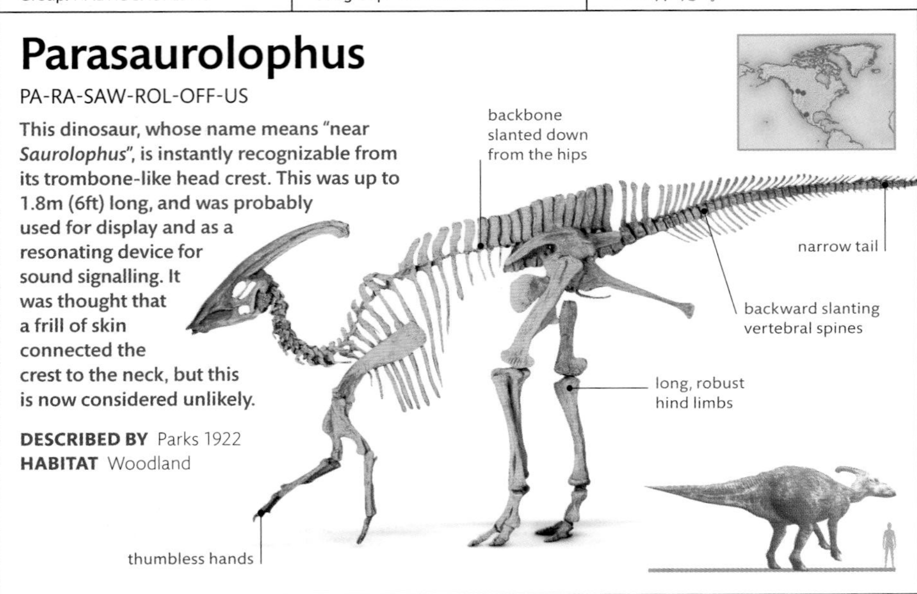

Length: 10m (33ft)	Weight: ~4 tonnes (3⅞ tons)	Diet: Leaves, seeds, and pine needles

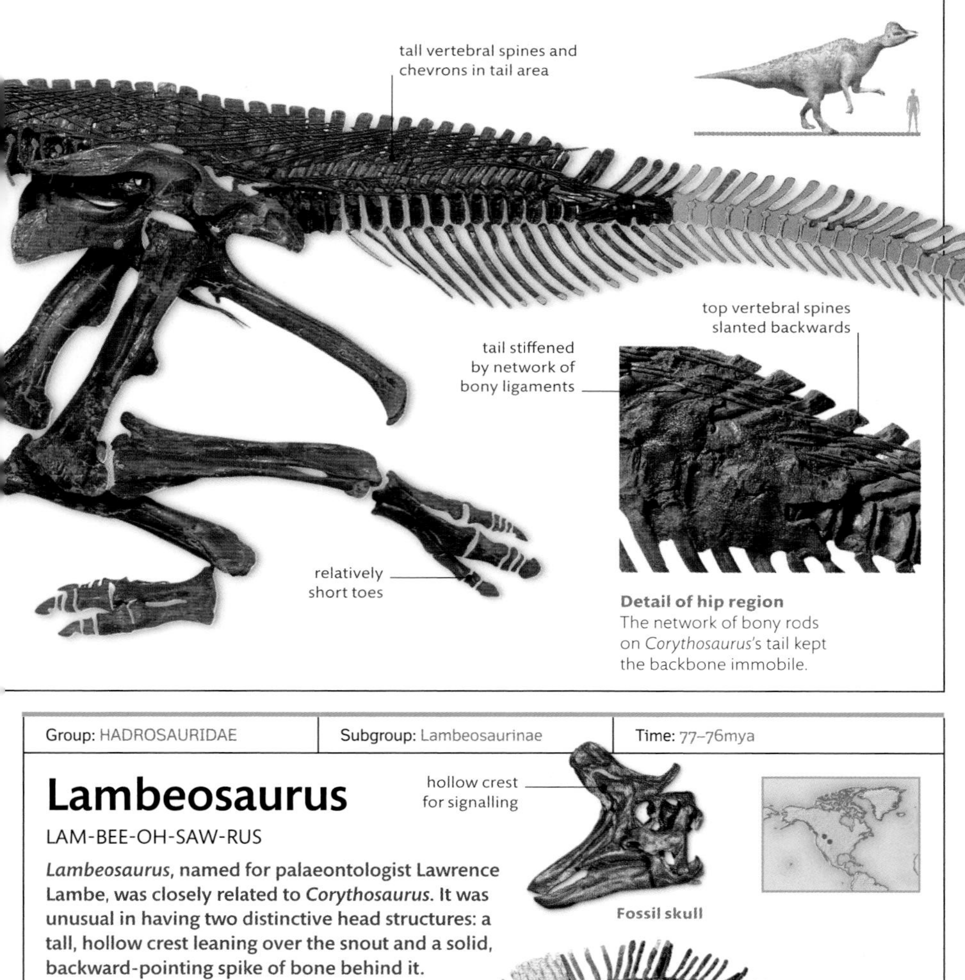

Detail of hip region
The network of bony rods on *Corythosaurus*'s tail kept the backbone immobile.

Group: HADROSAURIDAE	Subgroup: Lambeosaurinae	Time: 77–76mya

Lambeosaurus

LAM-BEE-OH-SAW-RUS

Lambeosaurus, named for palaeontologist Lawrence Lambe, was closely related to _Corythosaurus_. It was unusual in having two distinctive head structures: a tall, hollow crest leaning over the snout and a solid, backward-pointing spike of bone behind it. These were probably used for social recognition and signalling. Like other members of this family, it had a deep, narrow tail, held stiff and immobile. The animal moved around in large herds, browsing on low-growing vegetation on all-fours.

DESCRIBED BY
Parks 1923
HABITAT
Woodland

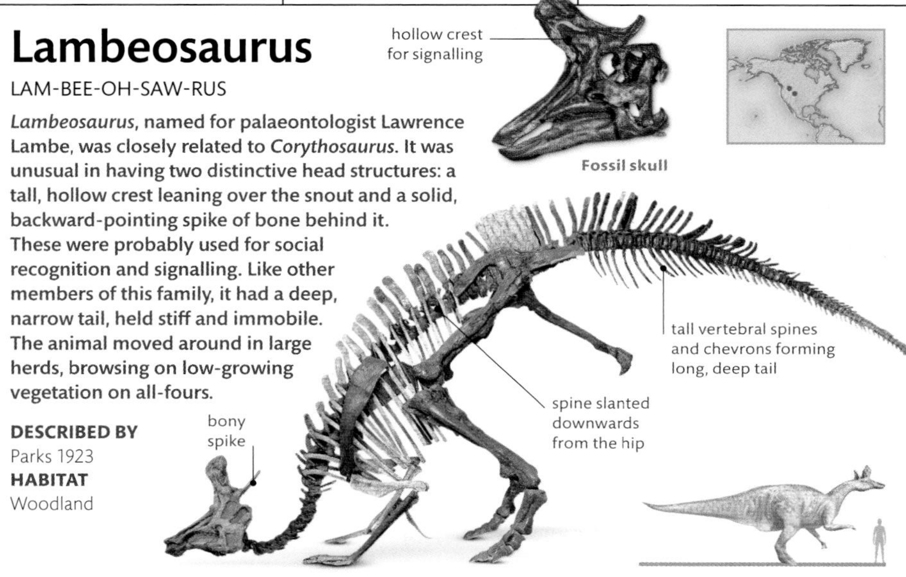

Fossil skull

Length: 9m (30ft)	Weight: ~4 tonnes (3⅞ tons)	Diet: Low-growing leaves, fruit, seeds

OTHER DIAPSIDS

AS FOR MOST OTHER groups of animals, the Cretaceous was a peak in diversity for many of the diapsids.

Turtles returned to the sea, and some, such as *Archelon*, reached gigantic sizes. Lepidosaurs diversified, and in the Late Cretaceous the first snakes appeared. At around the same time, a close relative of snakes returned to the water and gave rise to the mosasaurs, which would become the top predators in the seas. They were able to do so partly because ichthyosaurs had declined in number, eventually going extinct prior to the Cretaceous-Paleogene extinction.

Pliosaurs and plesiosaurs remained top aquatic predators throughout the Early Cretaceous, and continued to specialize. While the pliosaurs remained short-necked and developed enormous skulls to take down large prey, the plesiosaurs elongated their necks, culminating in the elasmosaurids, some with more than 70 vertebrae in the neck.

Throughout the Early Cretaceous, pterodactyloid pterosaurs radiated into a wide array of highly specialized forms. Some, such as *Quetzalcoatlus* (see p.154), reached incredible sizes unparalleled by any other flying animals. It was believed that pterosaurs declined in diversity in the Late Cretaceous, with only large-bodied lineages surviving, possibly as a result of competition with birds. However, numerous new studies show that small pterosaurs survived, only going extinct with the dinosaurs at the end of the Cretaceous.

Group: PLESIOSAURIA	Subgroup: Pliosauroidea	Time: 125–100mya

Kronosaurus

CRO-NO-SAW-RUS

Kronosaurus ("Kronos lizard") was a giant marine reptile, classed as a pliosaur because of its enormous head, compact body, and short neck and tail. Fossilized stomach contents show that *Kronosaurus* lived rather like a modern shark, devouring anything that came its way. Its head alone was 3m (9¾ft) long, and its sharp, pointed teeth were about 25cm (10in) long. Its snout was long and triangular. It had two pairs of flippers, the rear ones being longer than the front pair. There may have been a fin at the top of the tail to help steer the animal. The body of *Kronosaurus* was kept stiff as it swam by tightly linked belly ribs (gastralia).

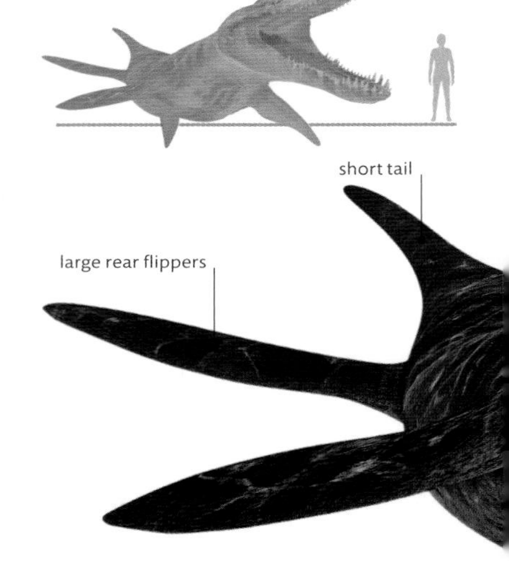

DESCRIBED BY Longman 1924
HABITAT Deep oceans

Length: 10m (33ft)	Weight: ~7 tonnes (7 tons)	Diet: Marine reptiles, fish, molluscs

Group: DIAPSIDA	Subgroup: Testudines	Time: 80–74mya

Archelon

AR-KAY-LON

This marine animal, whose name means "ruling turtle", was twice the length of modern species. It had a wide, flattened shell. This was formed from belly ribs (gastralia) that grew out from the body wall. The shell was composed of a leathery covering or horny plates over a framework of bony struts that may have been visible underneath.

DESCRIBED BY Wieland 1896
HABITAT Oceans

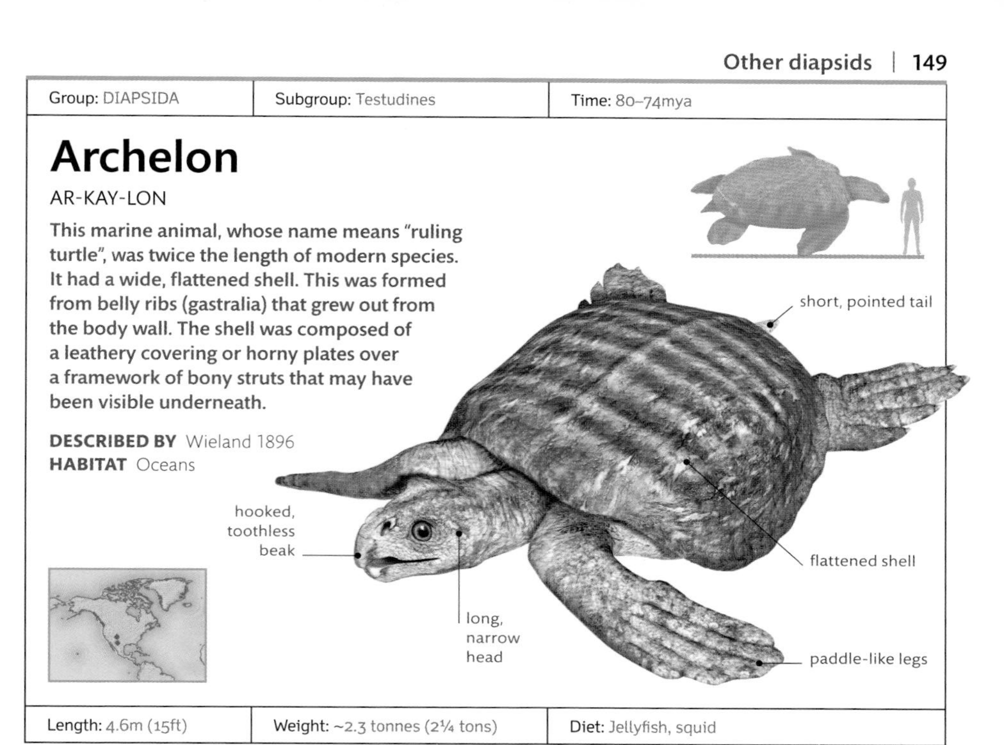

Length: 4.6m (15ft)	Weight: ~2.3 tonnes (2¼ tons)	Diet: Jellyfish, squid

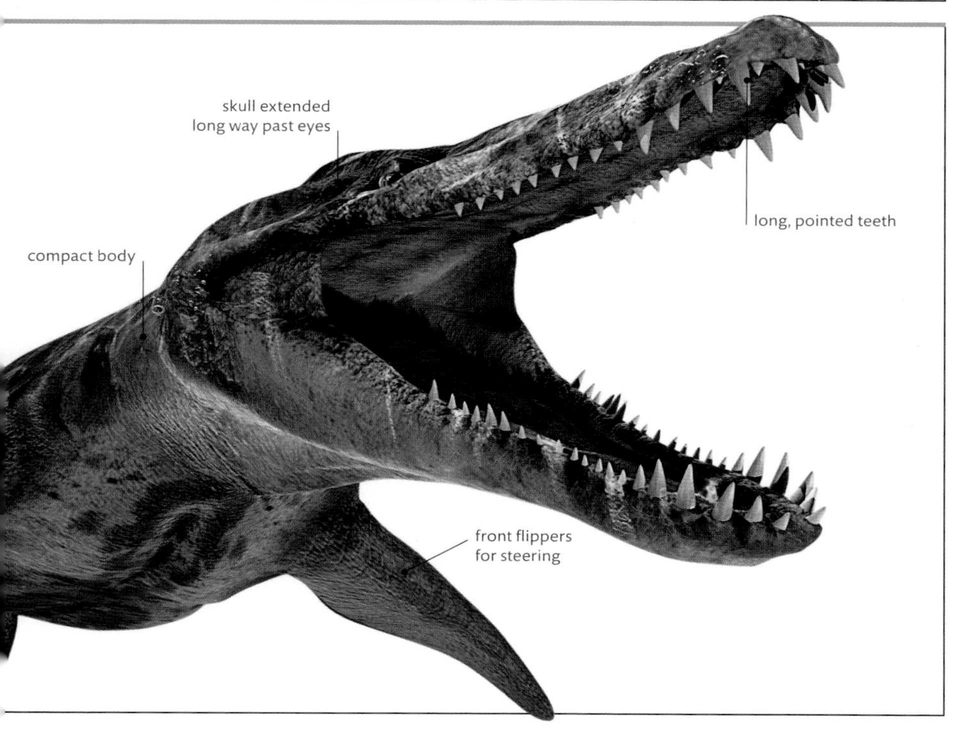

Group: PLESIOSAURIA	Subgroup: Plesiosauroidea	Time: 84–72mya

Elasmosaurus

EH-LAZ-MOH-SAW-RUS

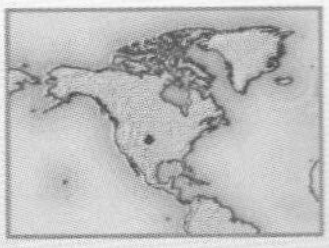

Elasmosaurus ("ribbon lizard") was the longest known plesiosaur. More than half of its body length consisted of the greatly elongated neck. This had 72 vertebrae, whereas earlier plesiosaurs had only 28. The length and structure of its neck meant that *Elasmosaurus* would have been able to snatch prey from a wide area around its head. It was earlier also thought to have held its neck high above water to spot prey. However, the idea is no longer supported as *Elasmosaurus*' centre of gravity would have limited its ability to raise its head too far out the water, except when its body was in contact with the sea bed. *Elasmosaurus*' body was similar to that of other plesiosaurs: it had four long, paddle-like flippers, a tiny head, sharp teeth in strong jaws, and a short, pointed tail.

DESCRIBED BY Cope 1868
HABITAT Oceans

greatly elongated neck

tiny head and mouth with small, sharp teeth

Length: 14m (46ft)	Weight: ~3 tonnes (3 tons)	Diet: Fish, squid, shellfish

Lepidotes

Elasmosaurus feasted on large fish such as *Lepidotes* (right) and many other forms of marine life that abounded in the Cretaceous oceans. *Lepidotes* was almost the length of a human and was itself a voracious predator of smaller marine creatures, such as shellfish, which it crunched with its strong teeth.

Group: SQUAMATA	Subgroup: Mosasauridae	Time: 93–66mya

Tylosaurus

TIE-LOW-SAW-RUS

This large sea lizard was one of the later members of the mosasaur family, and a fearsome predator. It had teeth on its palate bones as well as those in its long jaws. Its skull bones could move quite freely at the joints, allowing it to expand its jaws to swallow large prey. The snout was hard and bony at the tip (its name means "knob lizard"), leading some paleontologists to surmise that it may have rammed its prey. *Tylosaurus* probably swam by moving its deep, narrow tail in a side-to-side motion and steering with its large, wing-shaped flippers.

DESCRIBED BY Marsh 1872
HABITAT Shallow seas

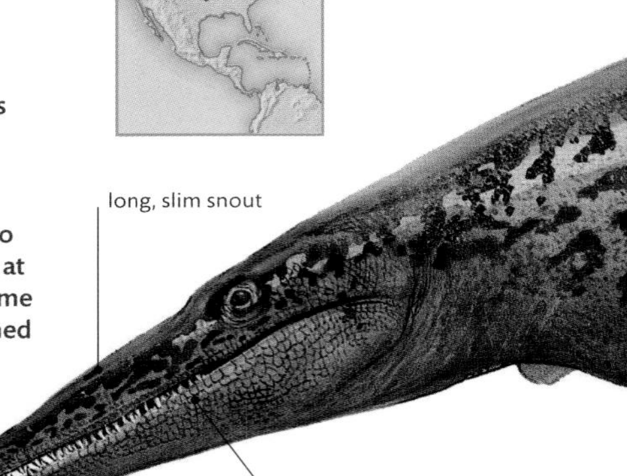

Length: 14m (46ft)	Weight: ~7 tonnes (6⅞ tons)	Diet: Turtles, fish, other mosasaurs

Group: SQUAMATA	Subgroup: Mosasauridae	Time: 83–66mya

Mosasaurus

MOH-SAH-SAW-RUS

This marine lizard, named for the Meuse River, was the first giant reptile ever to be named. It was one of the largest mosasaurs, with a slender, cylindrical barrel-like body, a long, powerful tail, and four long, paddle-like limbs. Its skull was strong, and the jaws contained many backward curving teeth capable of crushing and cutting. Healed injuries on fossil remains indicate that *Mosasaurus* lived a violent lifestyle. It is likely to have been a surface-swimming animal.

DESCRIBED BY
Conybeare 1822
HABITAT Oceans

long, flattened tail moved from side to side

streamlined body

paddle-like limbs

rosette of teeth

Length: 12.5–18m (40–59ft)	Weight: ~40 tonnes (40 tons)	Diet: Squid, fish, shellfish

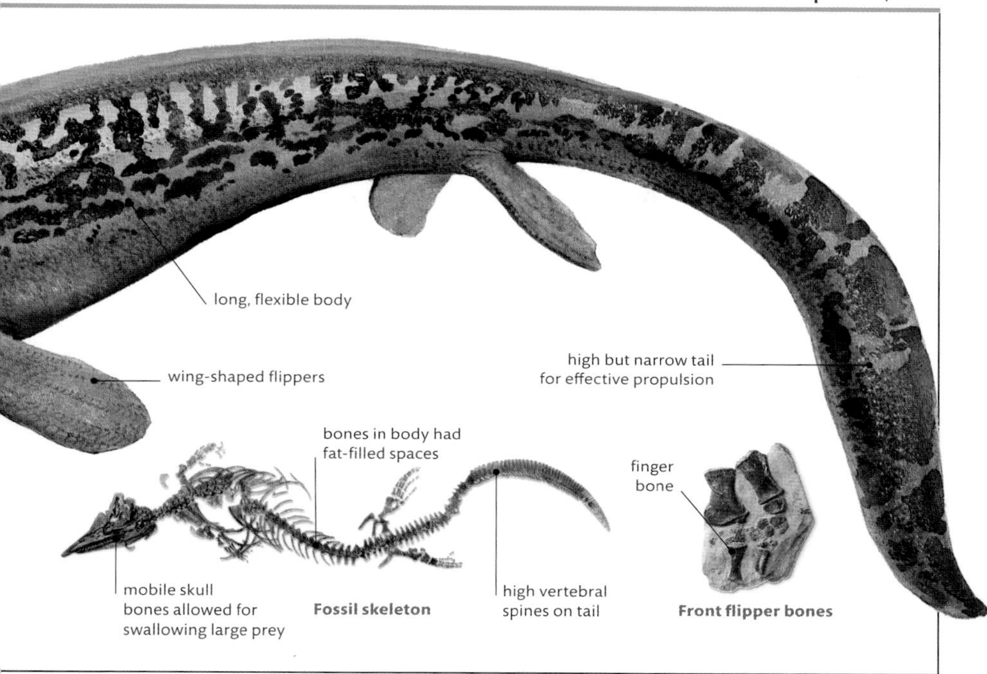

Group: SQUAMATA	Subgroup: Mosasauridae	Time: 72–66mya

Plotosaurus

PLOT-OH-SAW-RUS

Plotosaurus ("floating lizard") was an advanced mosasaur that greatly resembled an ichthyosaur. It had a slender, long skull, with large eyes and nasal cavities, and a deep, short, streamlined body. The tail was long and had a large, flat end formed from raised neural spines. It swam by moving its tail from side to side in a fish-like motion, and was probably among the fastest of the mosasaurs.

DESCRIBED BY Camp 1951
HABITAT Oceans

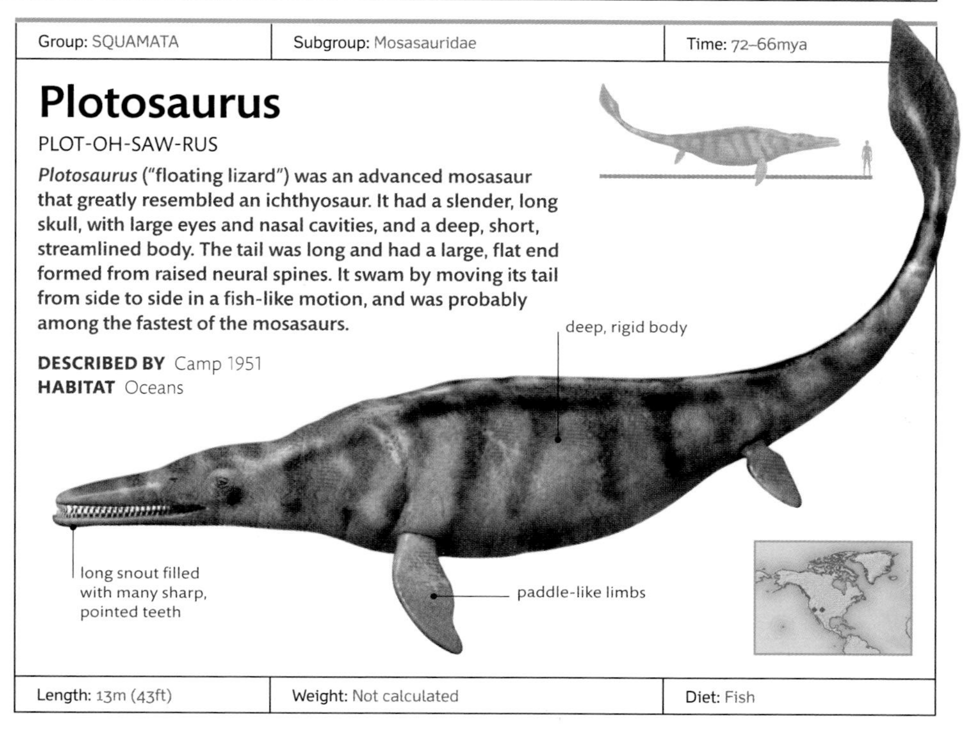

Length: 13m (43ft)	Weight: Not calculated	Diet: Fish

Group: PTEROSAURIA	Subgroup: Pterodactyloidea	Time: 125–100mya

Pterodaustro

TER-OH-DAS-TROH

The most remarkable feature of this large pterodactyl was its long, curved jaws. The lower jaw contained thousands of very thin teeth, which were probably used to sieve plankton. The upper teeth were used to comb out the lower teeth.

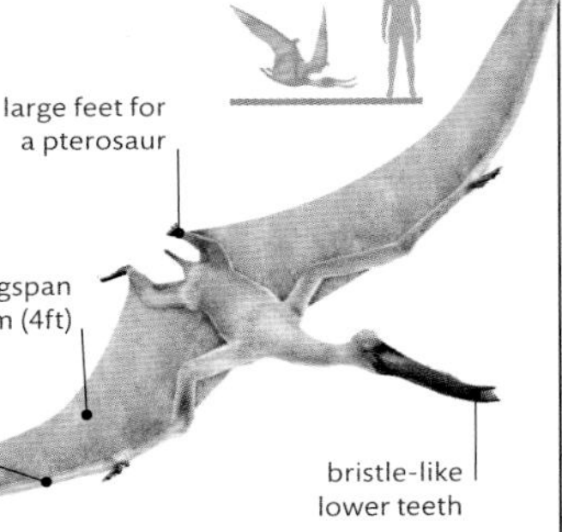

DESCRIBED BY
Bonaparte 1970
HABITAT
Seashores, lakes

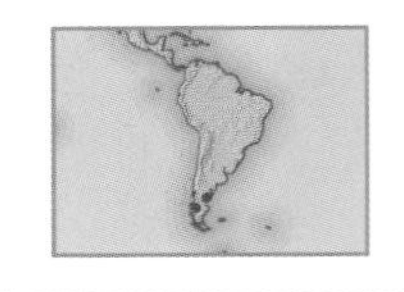

Length: 1.3m (4ft)	Weight: Not calculated	Diet: Plankton

Group: PTERODACTYLOIDEA	Subgroup: Pteranodontia	Time: 86–84mya

Pteranodon

TER-AN-OH-DON

***Pteranodon* was one of the largest pterodactyls. It probably flapped its wings to get off the ground and spent a lot of time soaring, actively flapping when necessary. The long crest on its head was probably used for sexual display.**

wingspan
of 7m (23ft)

long toothless
jaws possibly used
to scoop up fish

feet likely to
have been
webbed

DESCRIBED BY
Marsh 1876
HABITAT
Oceans, shores

Length: 1.8m (6ft)	Weight: ~16kg (35lb)	Diet: Fish

Group: PTERODACTYLOIDEA	Subgroup: Azhdarchoidea	Time: 68–66mya

Quetzalcoatlus

KWET-ZAL-KOH-AT-LUS

Remains of *Quetzalcoatlus*, indicating a wingspan of 11m (36ft) show it to have been the largest flying vertebrate known to date. Its wings were narrow and relatively inflexible. There was a small bony crest on the head.

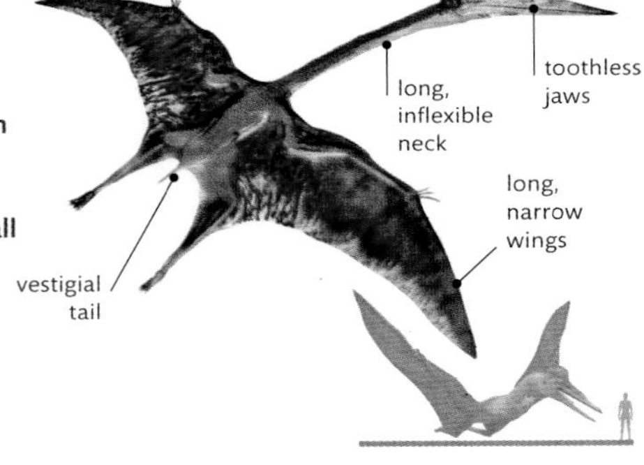

DESCRIBED BY
Lawson 1975
HABITAT All

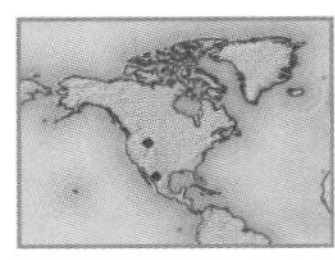

Length: 7.5m (25ft)	Weight: ~86kg (190lb)	Diet: Freshwater arthropods, carrion

Group: PTERODACTYLOIDEA	Subgroup: Pteranodontia	Time: 119–100mya

Tropeognathus

TROH-PEE-OH-NAY-THUS

Tropeognathus' most striking feature was the distinctive bulbous bony keel at the end of its long snout, which led to its name ("keel jaw"). This was probably patterned, and was used for signalling purposes or sexual display. This fish-eating pterosaur had a short tail and neck – typical features of its group. As in other pterosaurs, the wings were formed from thin membranous skin stretched between the elongated fourth fingers and the ankles. The body may have been covered in dense hair.

DESCRIBED BY Wellnhofer 1987
HABITAT Oceans, shores

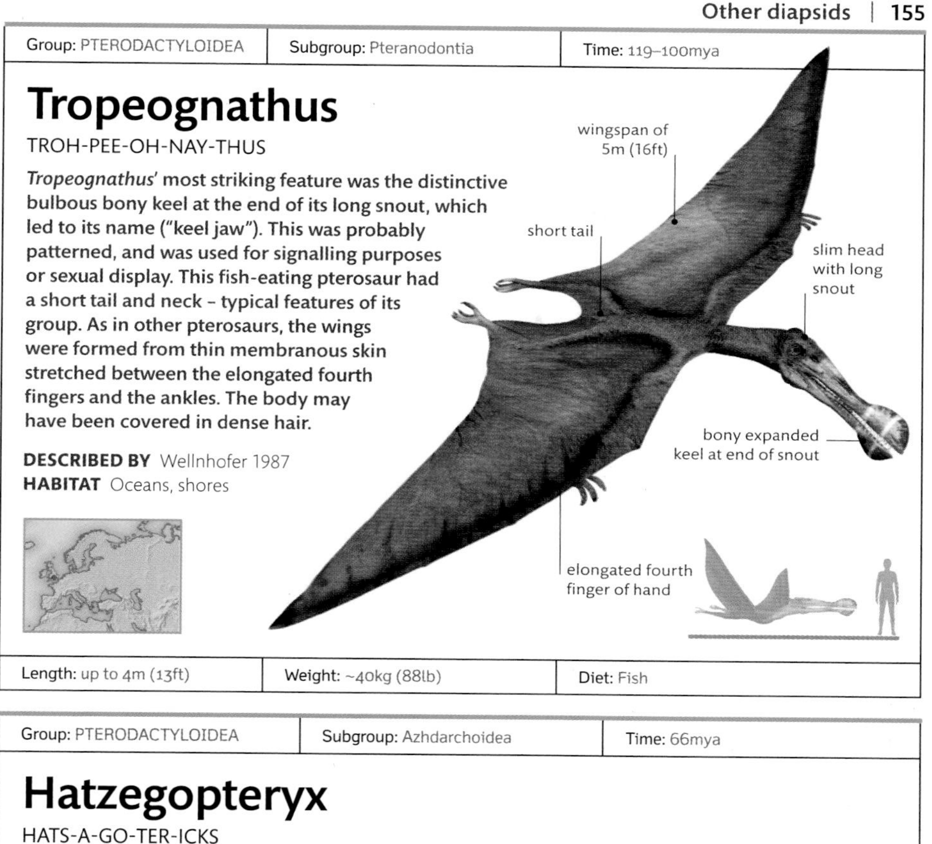

Length: up to 4m (13ft)	Weight: ~40kg (88lb)	Diet: Fish

Group: PTERODACTYLOIDEA	Subgroup: Azhdarchoidea	Time: 66mya

Hatzegopteryx

HATS-A-GO-TER-ICKS

Hatzegopteryx ("Haţeg Basin wing") was one of the largest flying animals of all time, with a wingspan of nearly 12m (39ft). Unlike other pterosaurs, which were lightly built, *Hatzegopteryx* had bulky bones and a short neck, suggesting that it would have been a fearsome, powerful predator. It was probably the apex predator of the Haţeg Basin, which at the time was an isolated island with no large land predators.

DESCRIBED BY Buffetaut et al. 2002
HABITAT Offshore islands

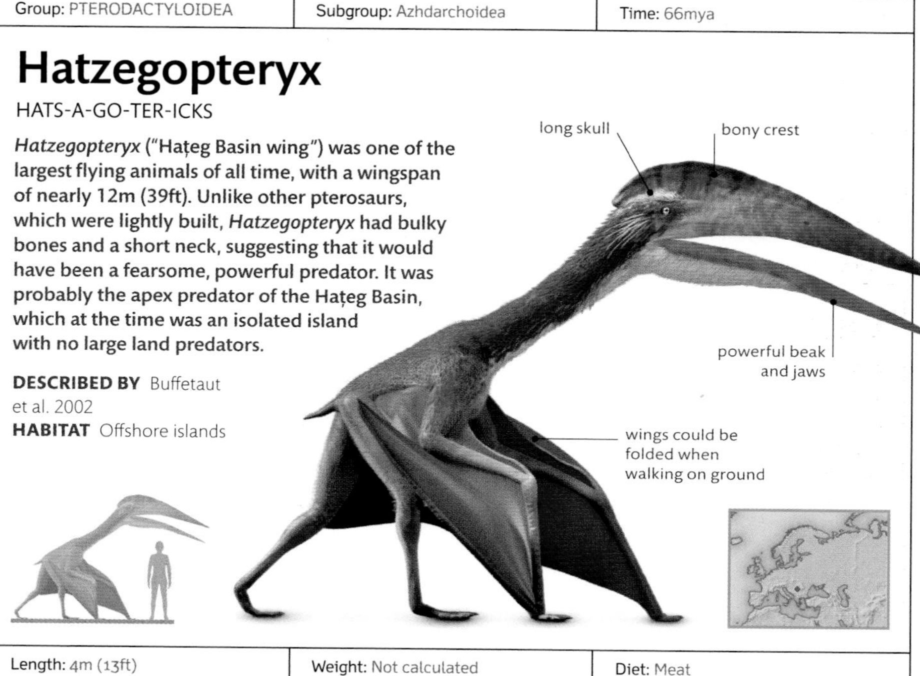

Length: 4m (13ft)	Weight: Not calculated	Diet: Meat

MAMMALS

AS IN THE JURASSIC, the mammals of the Cretaceous were relatively small but ecologically diverse. The multituberculates, which had evolved in the Jurassic, became globally widespread and abundant. This was enabled by their peculiar, efficient chewing apparatus - rodent-like incisors, and a large, blade-like premolar, with several rectangular molars designed for grinding. Triconodontids, another early-branching mammal group, were also successful, and some, such as the badger-sized *Repenomamus*, even preyed on small dinosaurs.

The only mammals that survive today, however, are from lineages that converged independently upon a key adaptation - complex molars that could both crush and shear in the same bite. One of these groups, from the southern hemisphere, gave rise to the monotremes, represented today by the platypus and the echidnas. The other group, from the northern hemisphere, split into the eutherians (ancestors of the placentals, including humans) and metatherians (ancestors of marsupials).

In the mid-Cretaceous, eutherians and metatherians diversified together with the rise of angiosperms and their pollinators. Eutherians were relatively small, specialized components of ecosystems, contrasting their ubiquity today. Instead, metatherians such as *Didelphodon* were far more widespread in the Cretaceous than they are today, and were the dominant mammals, alongside multituberculates, in Late Cretaceous ecosystems of Laurasia.

Group: MAMMALIA	Subgroup: Eutheria	Time: 83–76mya

Zalambdalestes

ZAH-LAMB-DAL-ES-TEES

Zalambdalestes was a shrew-like mammal, with a long tail and powerful hind legs that were longer than the front ones. The legs had elongated foot bones, and the hands were small, with non-opposable fingers. This suggests that *Zalambdalestes* was probably not a tree dweller. The eyes were very large, and the snout turned up sharply at the end. The incisors were long and sharp. The recent discovery of an epipubic bone (present only in non-placental mammals) in a skeleton of *Zalambdalestes* means it is considered a transitional form of mammal.

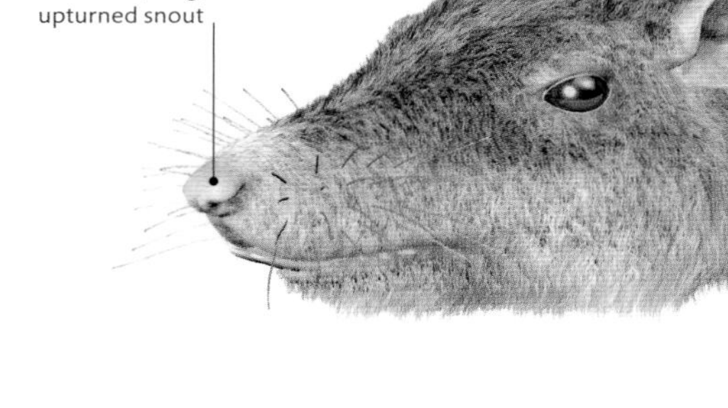

DESCRIBED BY Gregory and Simpson 1926
HABITAT Prairie

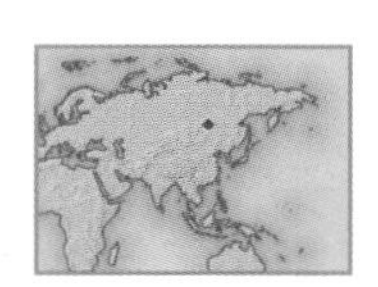

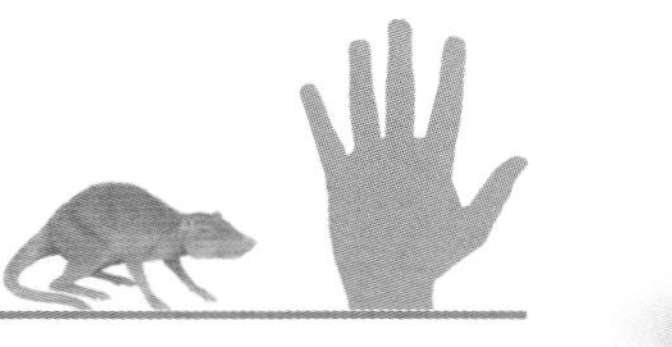

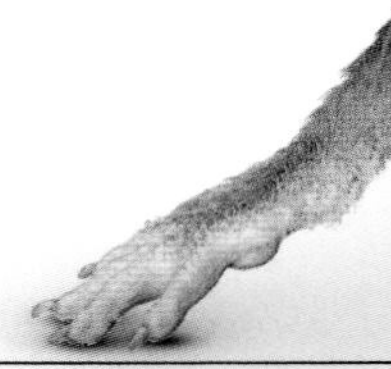

Length: 20cm (8in)	Weight: ~25g (1oz)	Diet: Insects

Group: MAMMALIA	Subgroup: Metatheria	Time: 72–66mya

Didelphodon

DIE-DEL-FOE-DON

Didelphodon **was one of the largest Mesozoic mammals. A heavily-built lower jaw and powerful teeth suggest that it had an incredibly strong bite for its size.** ***Didelphodon*** **could likely crush bones or shells with its bite, so it may have been both a predator and a scavenger, feeding on small prey, carrion, and molluscs. However, wear patterns on its teeth suggest that it supplemented its diet with insects and plants.**

DESCRIBED BY
Marsh 1889
HABITAT Swamps, floodplains, and riverbanks

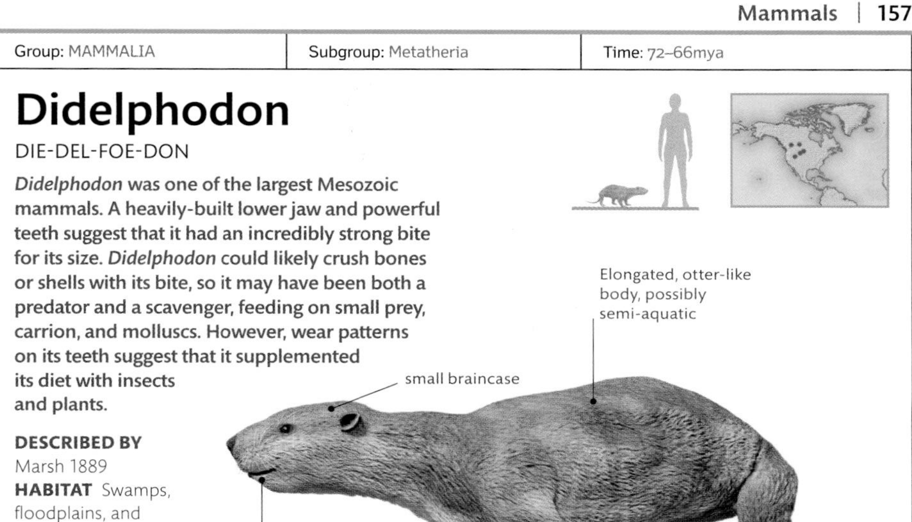

Length: 1m (3ft)	Weight: ~5kg (11lb)	Diet: Small prey, carrion, and molluscs

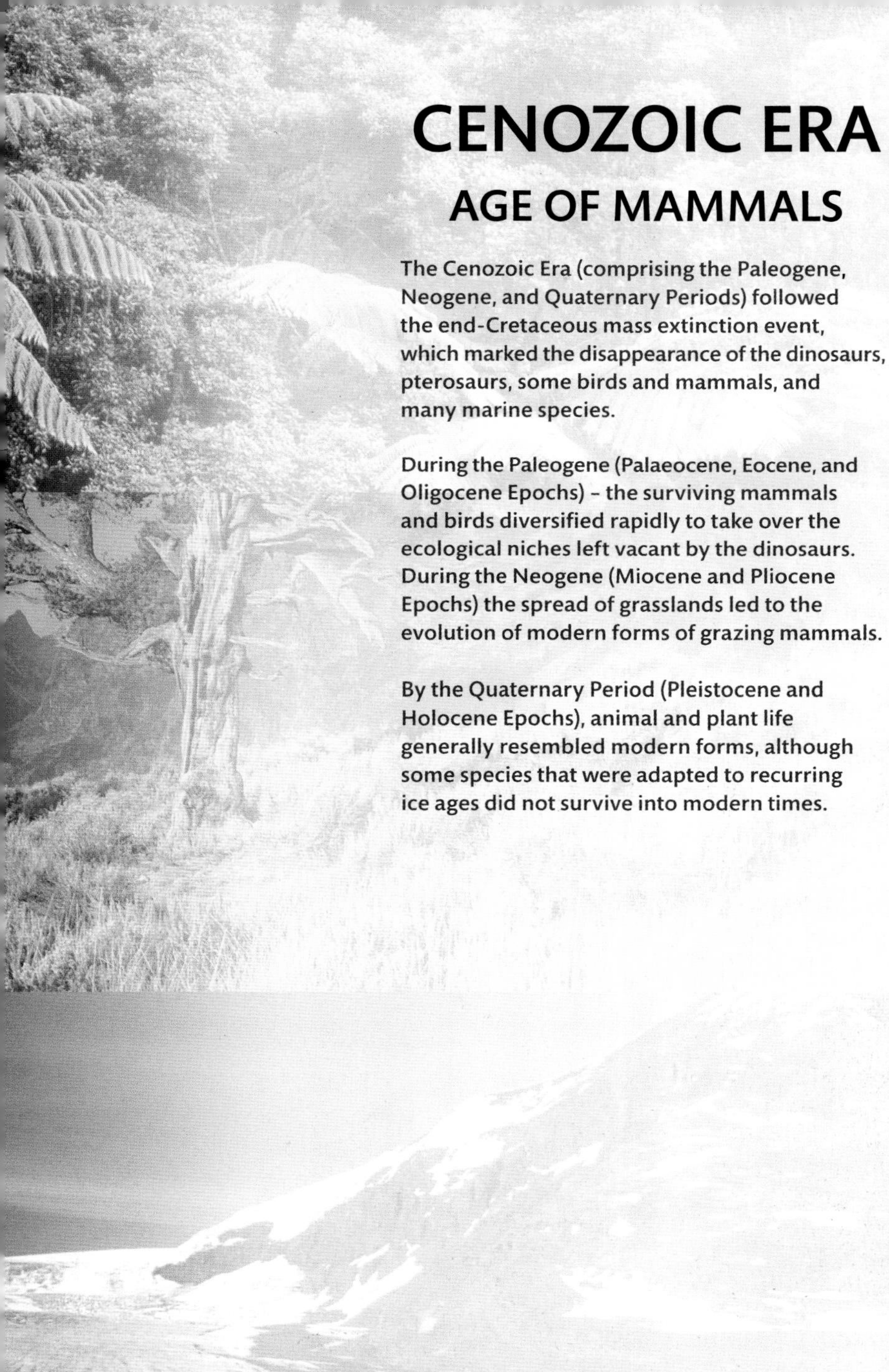

CENOZOIC ERA

AGE OF MAMMALS

The Cenozoic Era (comprising the Paleogene, Neogene, and Quaternary Periods) followed the end-Cretaceous mass extinction event, which marked the disappearance of the dinosaurs, pterosaurs, some birds and mammals, and many marine species.

During the Paleogene (Palaeocene, Eocene, and Oligocene Epochs) - the surviving mammals and birds diversified rapidly to take over the ecological niches left vacant by the dinosaurs. During the Neogene (Miocene and Pliocene Epochs) the spread of grasslands led to the evolution of modern forms of grazing mammals.

By the Quaternary Period (Pleistocene and Holocene Epochs), animal and plant life generally resembled modern forms, although some species that were adapted to recurring ice ages did not survive into modern times.

PALEOGENE PERIOD

66–23 MILLION YEARS AGO

THE PALEOGENE, including the Palaeocene, Eocene, and Oligocene epochs, was the beginning of the "Age of Mammals". Many ecological niches had been left vacant by the end-Cretaceous mass extinction of dinosaurs, and mammal and bird populations evolved and expanded rapidly to fill the gaps. Climates worldwide were warm or hot, with high rainfall. Vast areas of swampy forest and tropical rainforest developed. Towards the end of the Eocene, a large ice cap formed over the South Pole. This resulted in a fall in sea levels, and the climate became much cooler. Tropical forests disappeared in temperate regions and were replaced by woodlands dominated by deciduous and coniferous trees.

PALEOGENE LIFE

Despite the rapid evolution of mammals into many ecological niches, large herbivores only appeared at the end of the Palaeocene (about 53mya). There were some large mammalian carnivores, notably the mesonychids, as well as large predatory birds known as phorusrhacids. The reptile groups, including crocodiles and lizards, that had survived the end-Cretaceous extinction flourished in the warm, swampy conditions, as did smaller amphibians.

Icaronycteris
The first bats, such as *Icaronycteris* (right), first appeared in the Eocene. They filled the niche left by insectivorous pterosaurs, and like them, had wings formed from membranous skin.

wings supported on all fingers

feet thought to have been webbed for swimming

Ambulocetus
As well as diversifying on the land, mammals took to the oceans. The ancestors of whales, such as *Ambulocetus* (left), evolved from pig-like mammals. These primitive animals showed few adaptations to marine life.

4,600mya 4,000mya 3,000mya

PALEOGENE LANDMASSES

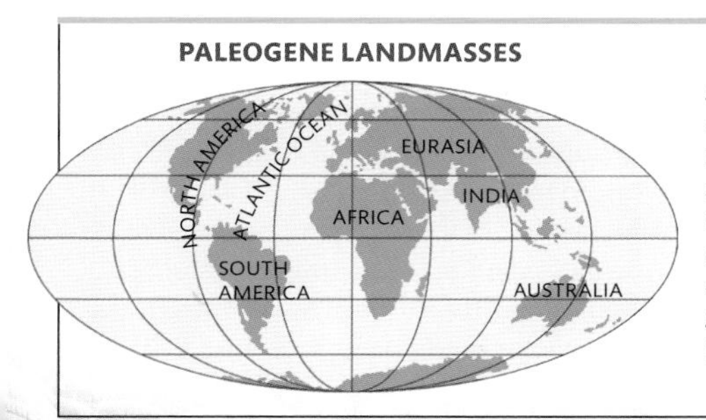

At the start of the Paleogene Period, the supercontinent of Gondwana was continuing to split up. South America was still an isolated island, and the Atlantic Ocean was becoming wider. By the end of the Eocene (see map) most of the continents were in similar positions to those they occupy today. India had begun to collide with Asia. Australia was moving north away from Antarctica, but had not yet reached its present-day position.

Uintatherium

Large herbivores, such as the uintatheres only appeared in the Eocene. They included the bizarre rhino-sized *Uintatherium* (right), which had three distinctive pairs of knobs on its head and tusk-like canine teeth.

2,000mya | 1,000mya | 500mya | 250mya | 0

MAMMALS

MAMMALS SUFFERED severe losses in the end-Cretaceous extinction, but they were able to recover rapidly. This may have been enabled by their generalized diet and small body sizes. Within a million years, they had already grown to larger sizes than throughout the entire Mesozoic, and took the top roles in ecosystems.

The balance of power shifted from the Cretaceous, and eutherians became more common than metatherians and multituberculates in most parts of the world. The major groups of placental mammals had probably originated prior to the end-Cretaceous extinction, but they radiated into many forms in the Palaeocene and early Eocene. These included modern groups, such as primates, carnivorans, and rodents, alongside several groups of uncertain relationships. By the Eocene, most of the placental groups living today had originated, and had spread into a broad array of lifestyles, including flying bats, fully aquatic whales, enormous herbivores, and fearsome carnivores. Metatherians also diversified during this time, with some becoming specialist fruit-eaters and others becoming large hypercarnivores.

Temperatures peaked early in the Eocene, leading to widespread forests, but towards the end of the Eocene they cooled, resulting in a turnover of the fauna into the Oligocene. The climate continued to become drier and cooler throughout the Oligocene, which led to the spread of grasslands. Many ungulates became adapted for running, and multituberculates went extinct after more than 130 million years on Earth.

Group: AFROTHERIA	Subgroup: Proboscidea	Time: 37–30mya

Phiomia

FI-OHM-EE-A

This early elephant had two pairs of tusks set in very long upper and lower jaws. The tusks on the lower jaw were flattened and formed a shovel-shaped projection that was probably used to collect food or to scrape bark off trees. The tusks in the upper jaw were shorter and were probably used for fighting or for display. The upper lip was likely to have been drawn out into a short trunk. To reduce the weight of the large skull, the bones were filled with air spaces. The rest of *Phiomia's* body was very similar to that of a modern elephant – with column-like legs and a stocky body.

DESCRIBED BY Andrews and Beadnell 1902
HABITAT Plains and woodland

Length: 5m (16ft)	Weight: ~3 tonnes (3 tons)	Diet: Plants

Group: PLACENTALIA	Subgroup: Afrotheria	Time: 41–27mya

Arsinoitherium

AR-SIN-OH-IH-THEER-EE-UM

Arsinoitherium **is the best-known member of the embrithopods – large, rhino-like animals that were closely related to elephants. Its most notable feature were the two huge, conical horns that jutted out from its snout. These were composed of hollow bone and may have been covered in skin. Adult animals had pointed horns, while they were rounded in juveniles.**

DESCRIBED BY Beadnell 1902
HABITAT Forests near rivers

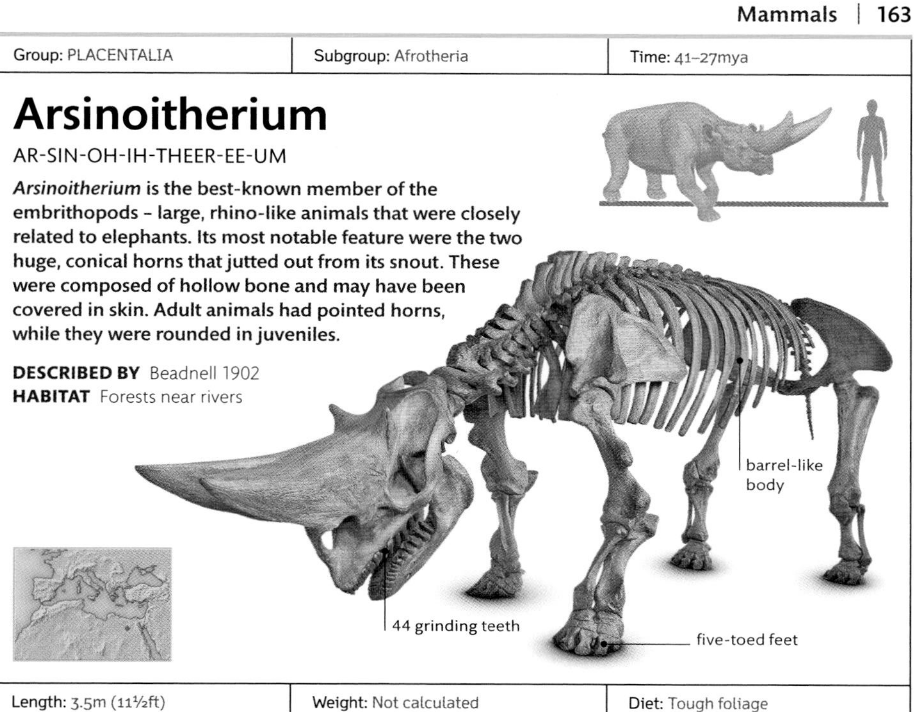

Length: 3.5m (11½ft)	Weight: Not calculated	Diet: Tough foliage

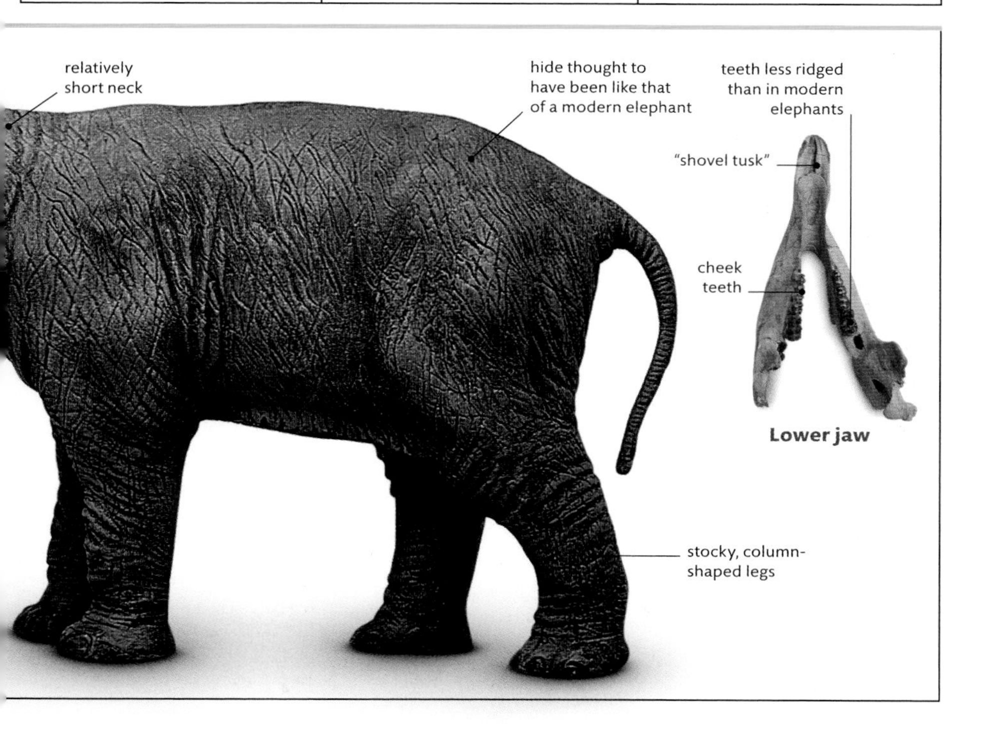

Group: AFROTHERIA	Subgroup: Proboscidea	Time: 41–34mya

Moeritherium

MEE-RI-THEER-EE-UM

This hippo-like animal, which was named after the ancient Greek name for the lake in Egypt where it was found (Moeris), is in the same group as elephants (Proboscidea). It had a low-slung, long body, with relatively short legs. Its nostrils were at the front of the skull, indicating that it did not have a trunk. The teeth were small, but two of the incisors formed small tusks. It was probably partly aquatic in its lifestyle, as is the modern hippopotamus.

DESCRIBED BY Andrews 1901
HABITAT Rivers and swamps

ears high on head

elongated, fleshy upper lip

Length: 3m (9¾ft)	Weight: ~200kg (440lb)	Diet: Water plants

Group: EUARCHONTOGLIRES	Subgroup: Primates	Time: 50–46mya

Notharctus

NOH-THARK-TUS

This animal is one of the best-known early members of the primate order. Looking very like a modern lemur, it showed many adaptations for a life spent in trees. Its eyes faced forwards, providing binocular vision for accurate judging of distances. Its legs and tail were long and strong. *Notharctus* had a type of grasping thumb that could be used for holding onto branches or food. Its fingers and toes were very long, which were also helpful for gripping. Its skull was short, and its long back was very flexible.

DESCRIBED BY Leidy 1870
HABITAT Forests

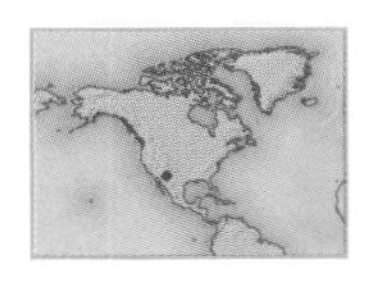

eyes set well forward

short, rounded skull

long, slim body

highly flexible spine consisting of small vertebrae

long, grasping fingers and thumb

hips set well back

long legs

Length: 40cm (16in)	Weight: ~4.5kg (10lb)	Diet: Leaves and fruit

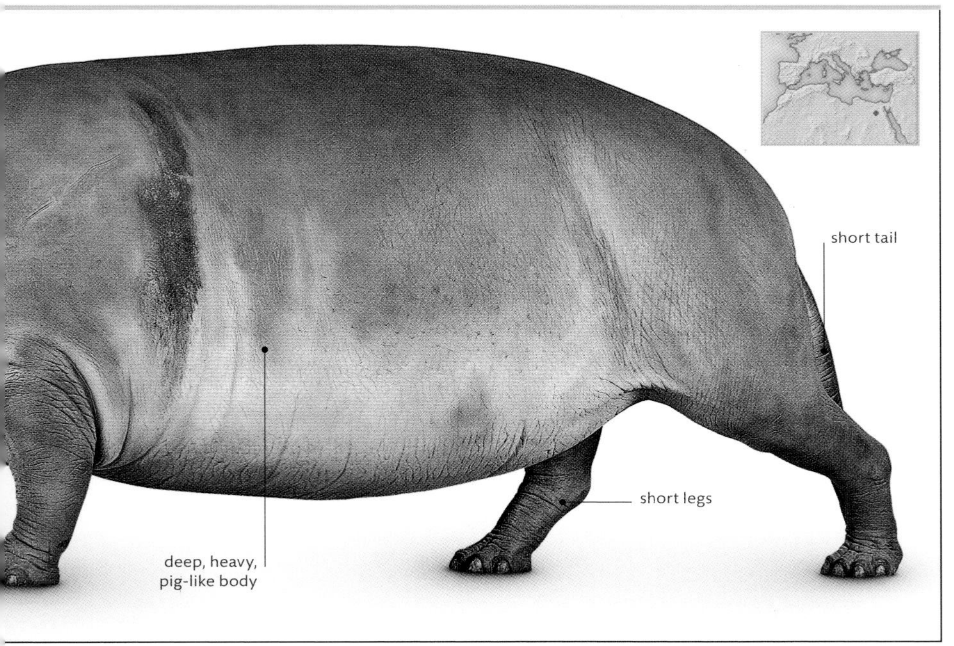

Group: LAURASIATHERIA	Subgroup: Chiroptera	Time: 52.5mya

Icaronycteris

IK-A-RON-IK-TER-IS

The oldest-known bat, *Icaronycteris* is surprisingly similar to modern forms. Analysis of its skull shows that it may have used a primitive form of echolocation. Its long tail was not joined to the legs by flaps of skin as in modern bats. The arrangement of the many teeth in the mouth indicate an insect diet. This has been confirmed by remains found in the stomach area of fossil *Icaronycteris*.

DESCRIBED BY Jepsen 1966
HABITAT Caves

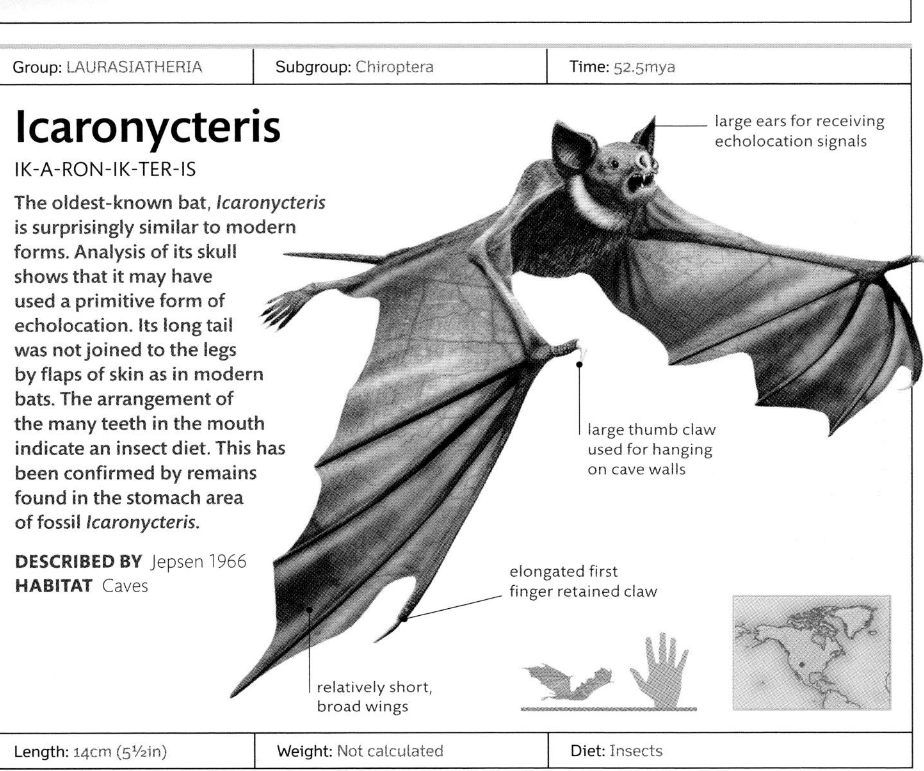

Length: 14cm (5½in)	Weight: Not calculated	Diet: Insects

Group: PLACENTALIA	Subgroup: Laurasiatheria	Time: 47–37mya

Mesonyx

MEE-ZON-ICKS

This mesonychid carnivore, whose name means "middle claw", had a wolf-like body and agile limbs that ended in five toes with small blunt claws. The long skull had a bony crest to which the large jaw muscles were anchored. This gave a powerful bite. The canine teeth were long and sharp, and the lower molars were thin and blade-like. Palaeontologists once thought that mesonychids were the ancestors of modern whales. However, based on new fossils and evidence from genetics, it is now clear that whales are more closely related to hippos.

wide mouth with powerful bite

DESCRIBED BY Cope 1871
HABITAT Scrubland, open woodland

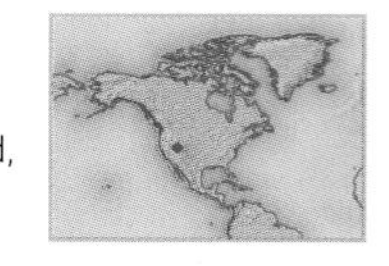

Length: 1.8m (6ft)	Weight: ~20–55kg (45–120lb)	Diet: Meat, carrion, possibly plants

Group: LAURASIATHERIA	Subgroup: Meridiungulata	Time: 40–37mya

Didolodus

DIE-DOH-LOH-DUS

The classification of *Didolodus* is still debated by palaeontologists. Some place it with early browsing and rooting ungulates (hoofed animals); others as a primitive litoptern (a now extinct group of hoofed mammals). Little is known of *Didolodus*. Its teeth were very similar to those of the earliest hoofed mammals, and so it may have resembled them in some respects. It was a browsing animal, with a long tail, and slim legs ending in five-toed feet.

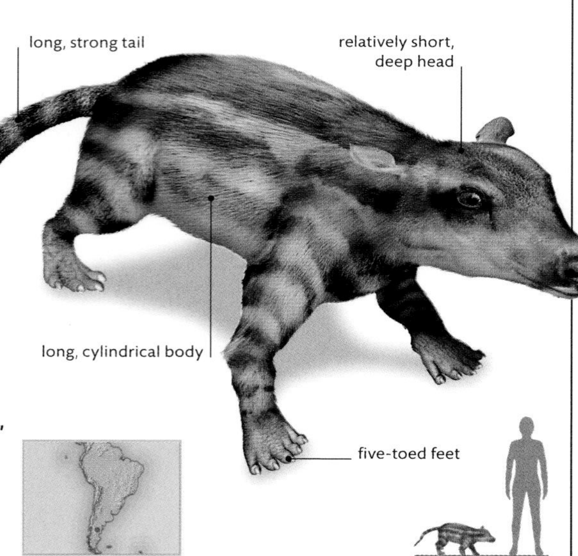

DESCRIBED BY Ameghino 1897
HABITAT Forests

Length: 60cm (24in)	Weight: ~15–20kg (33–44lb)	Diet: Leaves

Group: PLACENTALIA	Subgroup: Laurasiatheria	Time: 56–34mya

Uintatherium

YOU-IN-TAH-THEER-EE-UM

At the time when *Uintatherium* ("Uinta beast") existed, it was one of the largest land mammals. The size of a modern rhinoceros, it was heavy limbed, with massive bones. It had three pairs of horns on its face, which varied in size. The largest horns were at the back of the head. All of them seem to have been covered in skin and possibly hair. Males appear to have had larger horns than females. It is possible that the males fought using their horns and tusk-like canine teeth. The barrel-shaped body was carried on column-like legs. *Uintatherium* walked only on its toes, which were very short, as were the other "hand" and "foot" bones. Its elephant-like feet were designed for weight-bearing and walking on dry ground. *Uintatherium* had a very small brain for its skull size compared with modern hoofed mammals.

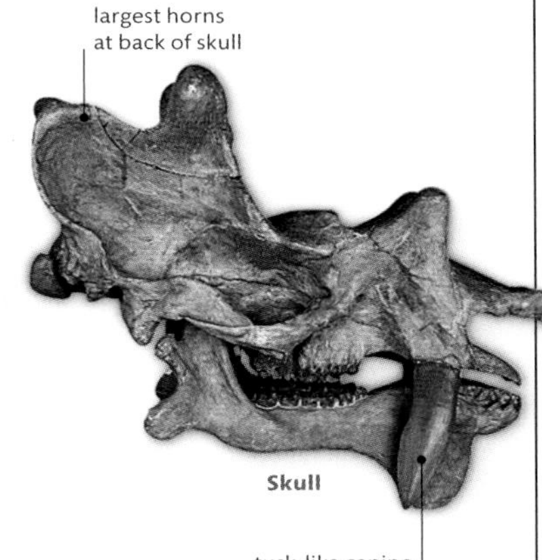

Skull

DESCRIBED BY Leidy 1872
HABITAT Forests

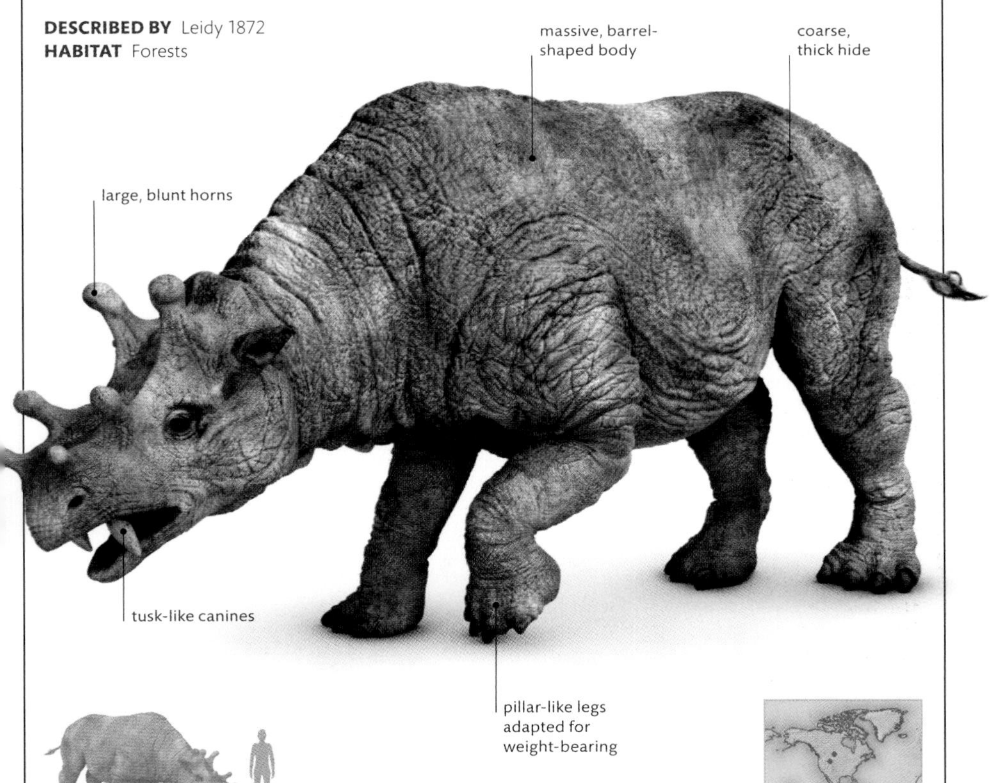

Length: 3.5m (11ft)	Weight: ~2 tonnes (2 tons)	Diet: Leaves, fruit, water plants

Group: LAURASIATHERIA	Subgroup: Perissodactyla	Time: 59–41mya

Phenacodus

FEN-A-COAD-US

Phenacodus was one of the earliest ungulate (hoofed) mammals. It was relatively small and lightly built, with limbs ending in five toes. The middle toe was the largest, and the weight of the body was mainly supported on this and the two adjoining toes. These probably had short, blunt hoof-like claws. The head was small, with a proportionately small brain. The jaws held 44 teeth. The back was arched, and the tail was long and powerful. *Phenacodus* probably lived in herds, and it is possible that it was carnivorous or insectivorous as well as herbivorous.

DESCRIBED BY Cope 1873
HABITAT Forests

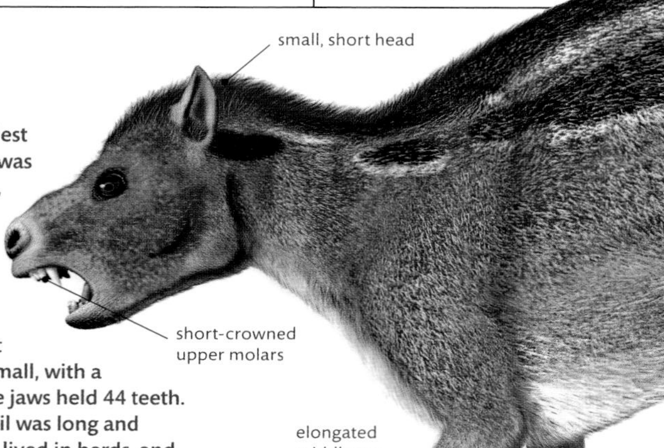

Length: 1.2m (4ft)	Weight: ~10kg (22lb)	Diet: Leaves; possibly tubers and insects

Group: LAURASIATHERIA	Subgroup: Perissodactyla	Time: 33–23mya

Paraceratherium

PAR-A-SER-A-THEER-EE-UM

This rhino relative was the largest land mammal known to have lived. Standing 5.25m (17ft) high at the shoulder, its head alone was over a metre (4ft) long. Hollowed out vertebrae in the back kept the animal's weight down. The powerful legs were long and slim, and there were three toes on each foot. Despite its long legs, it is now thought unlikely that *Paraceratherium* was a fast runner. It seems to have had a flexible upper lip, allowing it to browse on leaves from trees as modern giraffes do.

DESCRIBED BY
Forster-Cooper 1911
HABITAT Open woodland

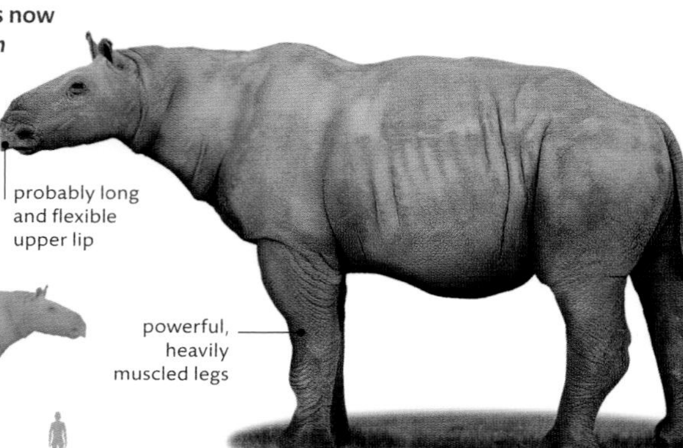

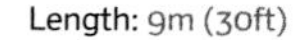

Length: 9m (30ft)	Weight: ~15 tonnes (14¾ tons)	Diet: Leaves and twigs

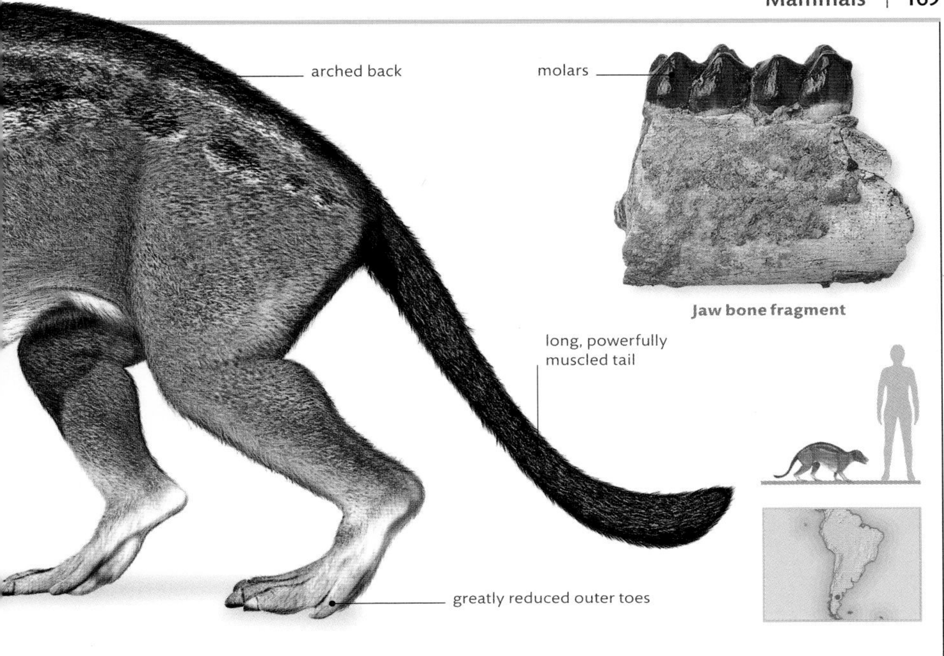

Jaw bone fragment

Group: LAURASIATHERIA	Subgroup: Perissodactyla	Time: 41–27mya

Mesohippus

ME-ZO-HIP-PUS

***Mesohippus* ("middle horse") had features that seem to have been an evolutionary response to more open environments: its legs were longer than those of the earlier "dawn horse" *Eohippus*, and it had lost a toe. Of the three remaining toes, the middle bore most of the animal's weight. These features increased its speed. Its teeth grew larger, increasing their surface area for chewing. The jaw was shallow and the head was quite long and pointed. The eyes were set relatively far apart and far back on the head.**

DESCRIBED BY
Marsh 1875
HABITAT
Open grassland

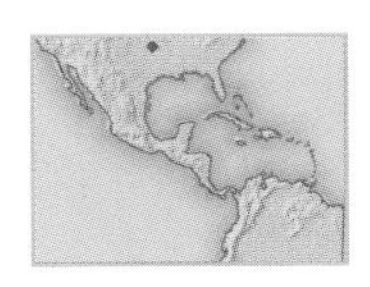

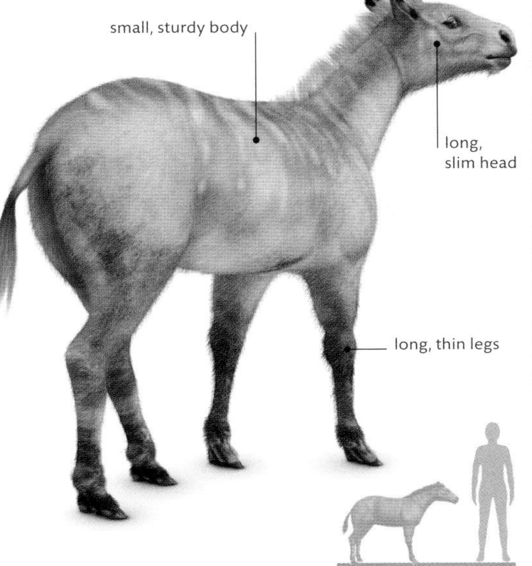

Length: 1.2m (4ft)	Weight: ~90kg (200lb)	Diet: Foliage

Group: CETARTIODACTYLA	Subgroup: Cetacea	Time: 48–41mya

Ambulocetus

AM-BYU-LOH-SEE-TUS

Ambulocetus ("walking whale") is thought to be one of the most primitive cetaceans (whales). Its teeth, skull, and ear bones show features unique to this group. As its name suggests, it probably spent most of its time on land, although it may have been rather clumsy. Its feet and hands may have been webbed, and its powerful hindlimbs would have made it a strong swimmer. It is thought to have hunted by ambushing its prey in water as modern crocodiles do.

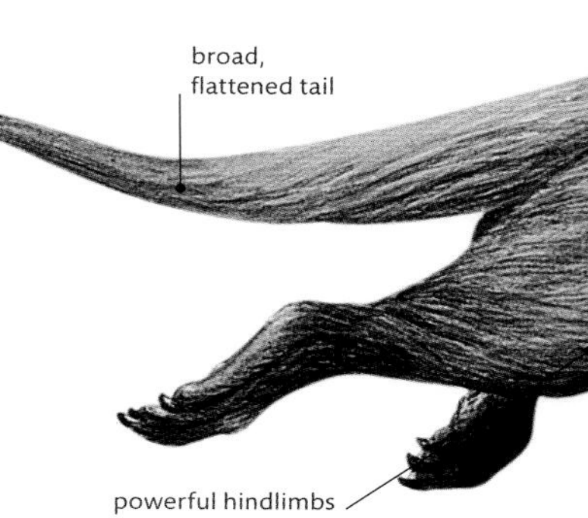

DESCRIBED BY Thewissen et al. 1994
HABITAT Estuaries

Length: 3m (9¾ft)	Weight: ~295kg (650lb)	Diet: Fish, mammals

Group: CETARTIODACTYLA	Subgroup: Cetacea	Time: 41–34mya

Basilosaurus

BASS-IL-OH-SAW-RUS

Despite being an early whale, *Basilosaurus* ("king of the lizards") looked so much like a mythical sea monster that its bones were at first thought to be those of an ancient sea reptile. It had a flexible, whale-like body, with forelimbs shaped like paddles. It had tiny hindlimbs that may have been used in mating. *Basilosaurus* probably swam by making undulating movements of its body. Its nostrils were high on its snout, but it did not have a blowhole. It was capable of hunting large fish and other marine mammals.

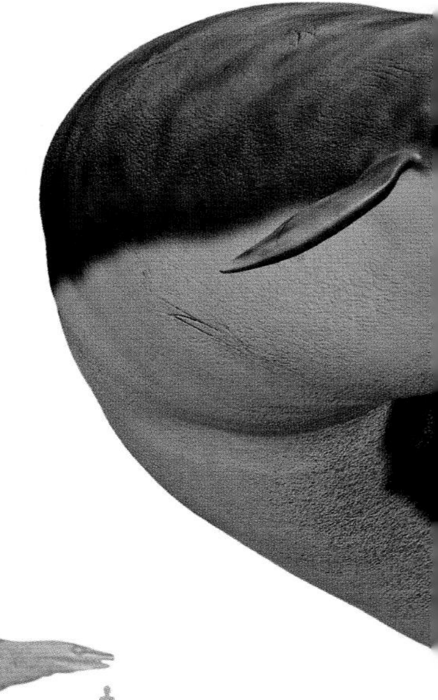

DESCRIBED BY Harlan 1834
HABITAT Tropical oceans

Length: 20–25m (66–82ft)	Weight: ~11 tonnes (10⅞ tons)	Diet: Large fish, squid, marine mammals

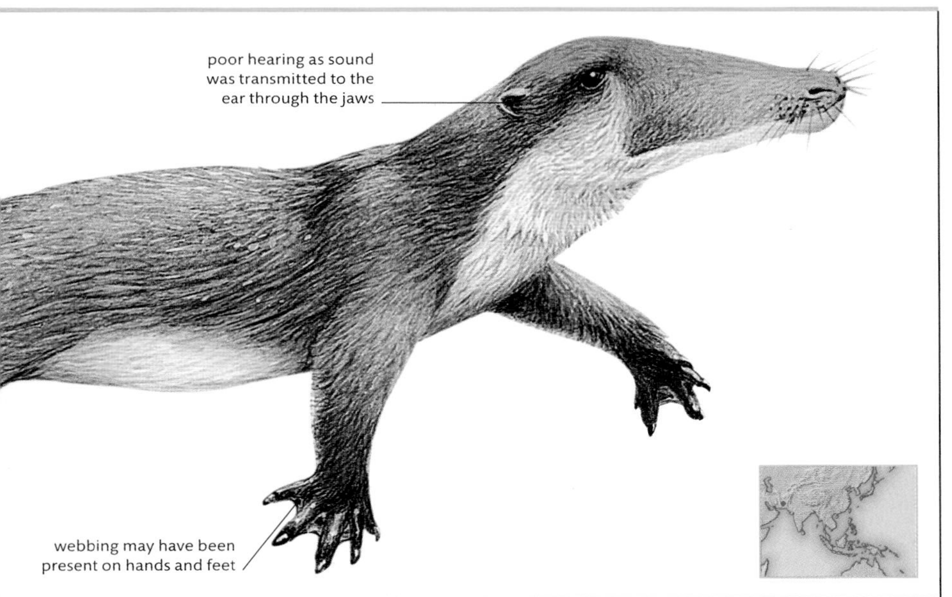
poor hearing as sound was transmitted to the ear through the jaws
webbing may have been present on hands and feet

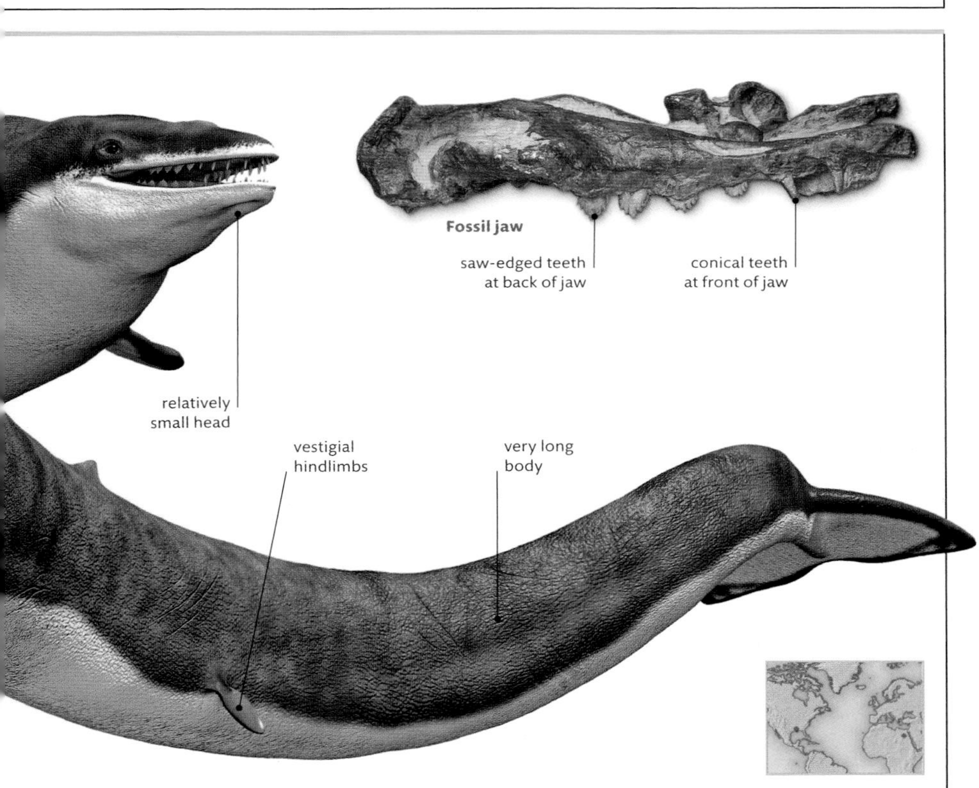
Fossil jaw
saw-edged teeth at back of jaw
conical teeth at front of jaw
relatively small head
vestigial hindlimbs
very long body

BIRDS

NEORNITHINE BIRDS survived the end-Cretaceous extinction, but the toothed enantiornithines and hesperornithids went extinct. It is debated whether any of the other archaic lineages of birds survived the extinction, but if they did, they went extinct shortly afterwards. The origination of different lineages of modern birds is also uncertain, although it is clear that the major branches diverged before the end-Cretaceous extinction.

In the early Paleogene, birds diversified and started to live different lifestyles, from small, tree-dwelling birds to larger terrestrial birds, and even early penguins. In the Palaeocene, ground-dwelling birds were among the largest animals on land and included herbivorous forms, such as *Gastornis* and the predatory phorusrhacids of South America.

The favourable warm climate and extensive development of forests in the early Eocene contributed to the spread and diversification of birds around the world. Late in the Eocene, the first songbirds evolved in Australia, dispersing around the world and exploding in diversity throughout the Oligocene. Meanwhile, the decline in temperature in the Oligocene led to a turnover in other species and the extinction of some lineages, such as the gastornithids. By the end of the Oligocene, songbirds were the most diverse lineage of birds, and today they still comprise 60 per cent of bird species.

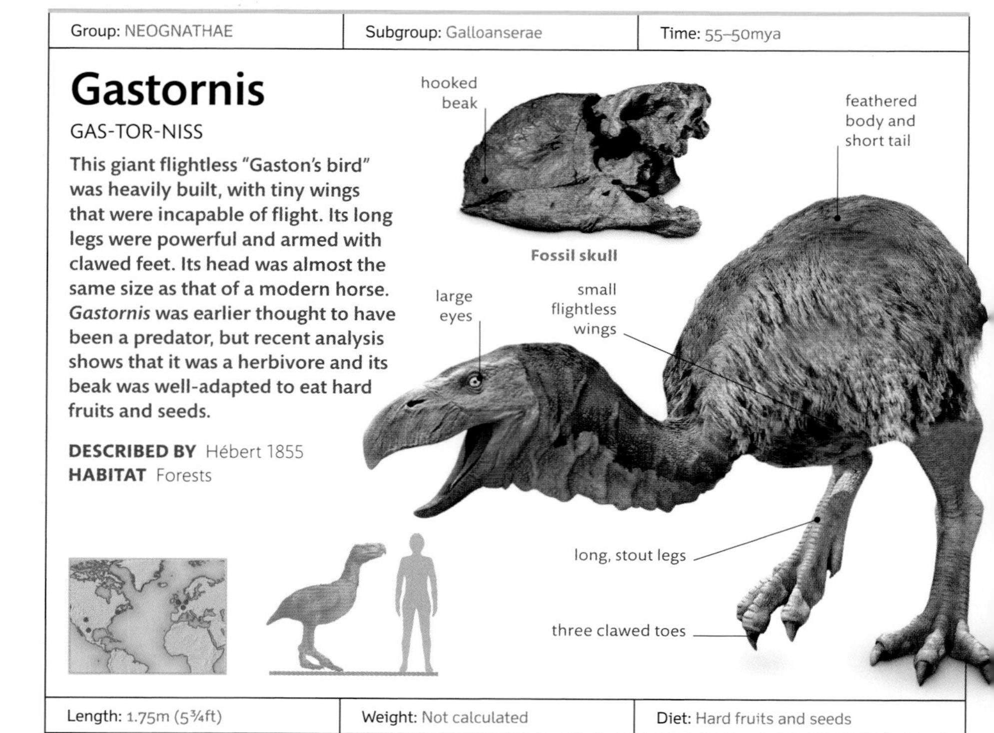

Group: NEOGNATHAE	Subgroup: Galloanserae	Time: 55–50mya

Gastornis

GAS-TOR-NISS

This giant flightless "Gaston's bird" was heavily built, with tiny wings that were incapable of flight. Its long legs were powerful and armed with clawed feet. Its head was almost the same size as that of a modern horse. *Gastornis* was earlier thought to have been a predator, but recent analysis shows that it was a herbivore and its beak was well-adapted to eat hard fruits and seeds.

DESCRIBED BY Hébert 1855
HABITAT Forests

Length: 1.75m (5¾ft)	Weight: Not calculated	Diet: Hard fruits and seeds

Group: NEOGNATHAE	Subgroup: Galloanserae	Time: 61–33mya

Presbyornis

PREZ-BEE-OAR-NIS

This primitive duck was so slenderly built that palaeontologists first thought it was a type of flamingo. Unlike modern ducks, *Presbyornis* had very long legs and presumably was a wader, rather than a swimmer. The relative length of the neck and legs of *Presbyornis* are consistent with a wading, bottom-feeding style. Although the legs were set relatively far back, it was unlikely to have been a diving bird. Its feet were large and webbed. *Presbyornis* was highly colonial, and seems to have gathered in large flocks to feed by the shores of lakes. Recent analysis of the jaw mechanism shows that it was unable to filter feed like modern ducks, and was therefore a mixed feeder, similar to the magpie goose (*Anseranas semipalmata*). Hundreds of *Presbyornis* fossils have been found, representing several species that range from 0.5m (1½ft) to 1.5m (5ft) in height.

DESCRIBED BY Wetmore 1926
HABITAT Lakeshores

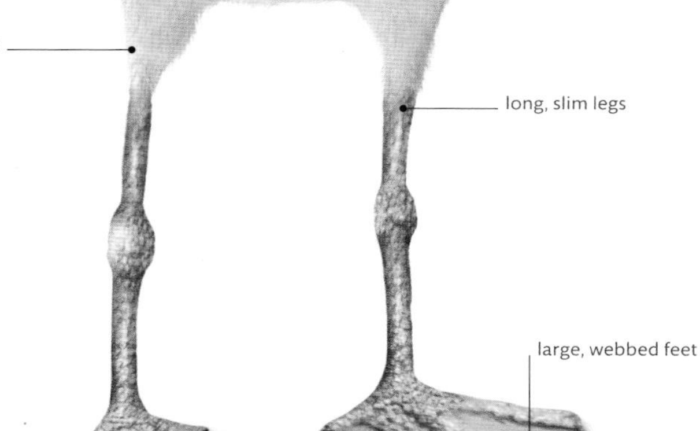

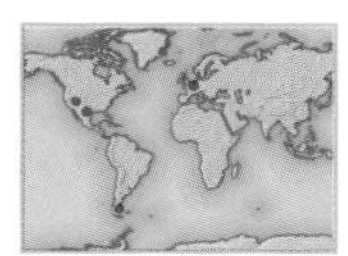

Length: 0.5–1.5m (1½–5ft)	Weight: Not calculated	Diet: Plankton, water plants

NEOGENE PERIOD

23–2.6 MILLION YEARS AGO

THE NEOGENE (the Miocene and Pliocene epochs) saw the continued evolution of mammals. The ice cap that had started to form at the South Pole in the Oligocene continued to expand. By the mid-Miocene, it covered the whole of Antarctica, further cooling the world's climate. These temperate conditions led to the development of huge expanses of grassland over Africa, Asia, Europe, and the Americas. The climate continued to cool into the Pliocene, when a large ice cap also formed at the North Pole. Most of the remaining forest areas had disappeared by the end of the Pliocene, but grasslands continued to spread.

NEOGENE LIFE

The animals of the Miocene and Pliocene were very similar to modern forms. Grasslands were filled with herds of horses, camels, elephants, and antelopes. Large carnivores hunted across the plains, and the first hominins evolved from primate ancestors. In the oceans, modern types of fish had appeared, and the primitive forms of whales had given way to more familiar, larger species.

reduced wings solely for display

sharp, hooked beak

Terrible birds

Large carnivorous birds were still some of the world's largest and most terrifying predators. The South American terror bird *Titanis* (left) hunted on open grassy plains. It moved into North America when the Panama Isthmus formed in the Pliocene.

4,600mya 4,000mya 3,000mya

NEOGENE LANDMASSES

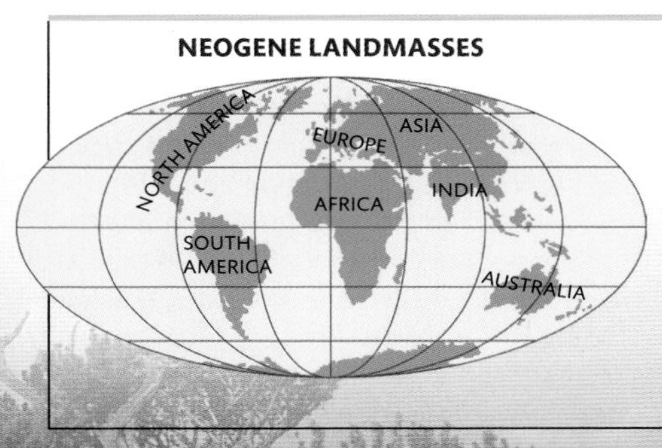

By the Neogene, the continents were nearly identical to today, including most major mountain ranges. Throughout the Triassic–Paleogene, Africa and Europe collided, forming the Alps. India's collision with Asia caused the Himalayas to rise about 40mya. The Rocky Mountains in North America were formed during the Cretaceous–Palaeocene, and the Andes in South America formed during the Triassic, uplifting throughout the Cretaceous. By the Pliocene, a land bridge linked North and South America. Australia continued to drift northwards.

Big cats
Many carnivores evolved large canine teeth. *Thylacosmilus* (above) was a cat-like marsupial found in Miocene and Pliocene South America.

Early elephants
By the beginning of the Miocene, early elephant forms, such as the horse-sized *Phiomia* (left) were dying out, and being replaced by larger forms. These, such as the shovel-tusked *Platybelodon*, would give rise to the mammoths of the Pleistocene and to modern elephant species.

MAMMALS

THE MAMMALS of the Miocene were a mix of archaic Paleogene lineages and early members of many modern lineages.

In Eurasia, ungulates (hoofed mammals) were common, and the perissodactyls, such as rhinoceroses, horses, and the chalicotheres, were far more diverse than today. Artiodactyls were represented by oreodonts, camels, and the omnivorous entelodonts. Early carnivorans like bears, dogs, and cats, were beginning to diversify, but the top predators, such as the saber-toothed nimravids and the "creodonts" were still archaic forms. Proboscideans, represented by the elephants today, were incredibly diverse and radiated in the Miocene as they dispersed from Africa. Whales reached a peak in diversity, and hominid apes began to evolve in Africa.

In South America, xenarthrans, such as the giant, armoured glyptodonts, flourished alongside a unique group of hooved animals called meridiungulates - distant cousins of the perissodactyls. Primates and rodents, which had somehow reached South America in the Eocene by rafting across the Atlantic Ocean, diversified into their own unique forms.

The Pliocene saw the continued replacement of archaic lineages with modern ones. Towards the end of the Pliocene, North and South America became reconnected, leading to the Great American Biotic Interchange. The unique faunas mixed, resulting in extensive dispersal, restructuring of ecosystems, and many extinctions. Meanwhile, human ancestors spread out of Africa.

Group: MAMMALIA	Subgroup: Metatheria	Time: 5.3–2.6mya

Thylacosmilus

THIGH-LA-CO-SMILE-US

Thylacosmilus was a large carnivorous marsupial. Like the sabre-tooth cats, it sported long, stabbing upper canine teeth. However, in *Thylacosmilus*, these teeth grew continuously throughout life, and there were no incisor teeth in the lower jaw. Bony guards in the lower jaw protected the teeth when the jaws were closed. The neck and shoulders were strong and heavily muscled, which allowed for the teeth to be pulled back to disembowel carcasses. The anatomy of *Thylacosmilus* shows that it was not an active predator, but was most likely a scavenger that specialized in eating the internal organs of carrion.

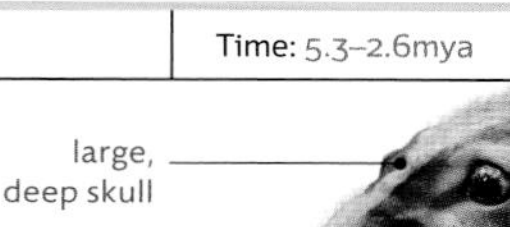

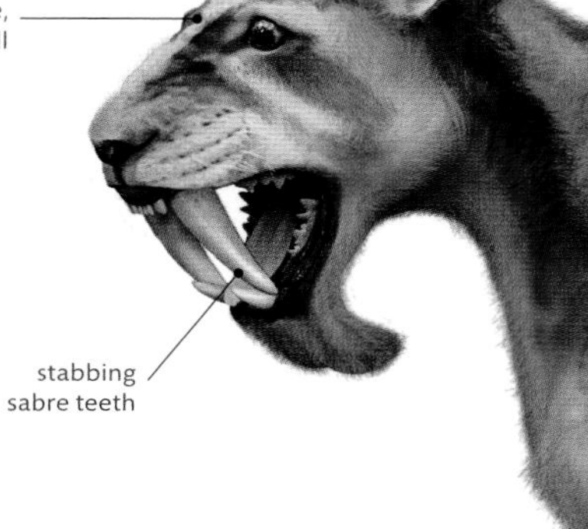

DESCRIBED BY Riggs 1933
HABITAT Plains

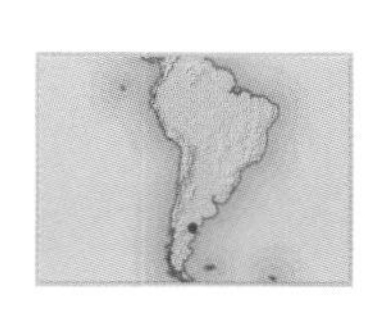

Length: 1.2m (4ft)	Weight: ~115kg (250lb)	Diet: Slow-moving, hoofed mammals

Group: MAMMALIA	Subgroup: Metatheria	Time: 18–16mya

Cladosictis

CLAD-OH-SICK-TIS

This carnivorous marsupial species was short lived. Some palaeontologists have suggested that it may have had an otter-like lifestyle – hunting fish in rivers – but it may also have eaten the eggs and young of land-living creatures. *Cladosictis* had a long, lightly built body, short limbs, and a dog-like skull. Its tail was long and thin. The teeth were similar to those of carnivorous placental mammals: there were sharp incisors at the front of the jaw, with pointed canines and shearing molars behind them.

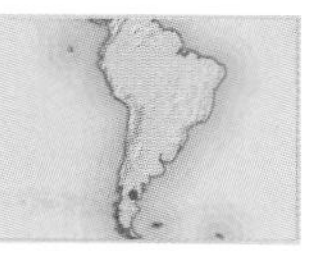

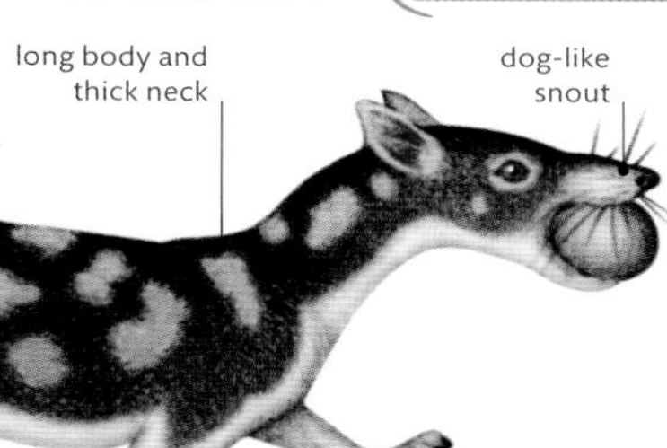

DESCRIBED BY
Ameghino 1887
HABITAT Woodland

Length: 80cm (2½ft)	Weight: ~3.5–8kg (8–18lb)	Diet: Small animals, perhaps fish and eggs

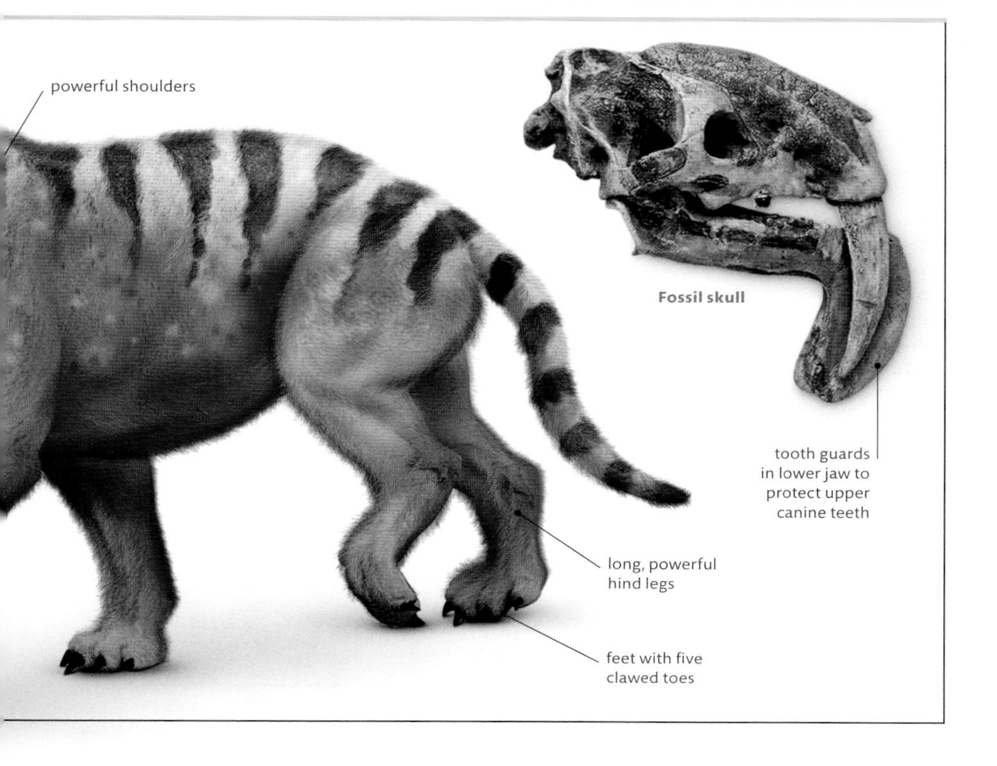

Fossil skull

Group: AFROTHERIA	Subgroup: Proboscidea	Time: 18–4mya

Gomphotherium

GOM-FO-THEER-EE-UM

Gomphotherium had a pair of upper jaw tusks for fighting and display, and long lower jaws that formed a food "shovel". It had a trunk as long as the lower tusks to help with feeding. Later species had fewer teeth, with more pronounced ridges for grinding.

DESCRIBED BY Burmeister 1837
HABITAT Grassland, marsh, and forest

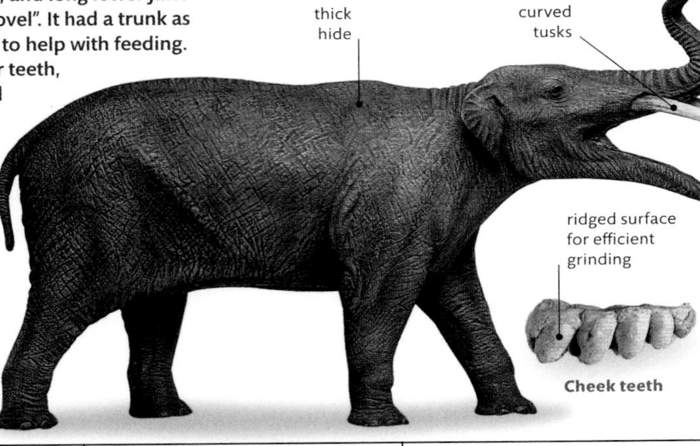

Length: 5m (16ft)	Weight: ~4–7 tonnes (3⅞–6⅞ tons)	Diet: Plants

Group: AFROTHERIA	Subgroup: Proboscidea	Time: 16–11mya

Platybelodon

PLAT-EE-BEL-OH-DON

The shovel arrangement of the lower tusks in this family of elephants reached extreme proportions in _Platybelodon_. The flattened lower tusks were very broad, with an indentation near the top for the upper pair of tusks to slot into while the mouth was closed. The cheek teeth were flat and there were sharper teeth at the front of the jaw. The trunk was very wide. _Platybelodon_ seems to have used its shovel-tusks to scoop up water weeds and other soft plants. The tail was relatively long. As a highly specialized feeder, it was vulnerable to environmental change, and was a short-lived genus.

DESCRIBED BY Borissiak 1928
HABITAT Wet prairies

Length: 6m (20ft)	Weight: ~4–5 tonnes (3⅞–4⅞ tons)	Diet: Soft water plants

Group: LAURASIATHERIA	Subgroup: Carnivora	Time: 20–2.6mya

Amphicyon

AM-FIE-SIGH-ON

The amphicyonids were a family of "bear-dogs", of which *Amphicyon* ("ambiguous dog") was a typical member. A powerful predator, it had well-muscled shoulders, a heavily built body, and strong legs. It may have lived a lifestyle similar to that of modern bears.

DESCRIBED BY Lartet 1836
HABITAT Plains

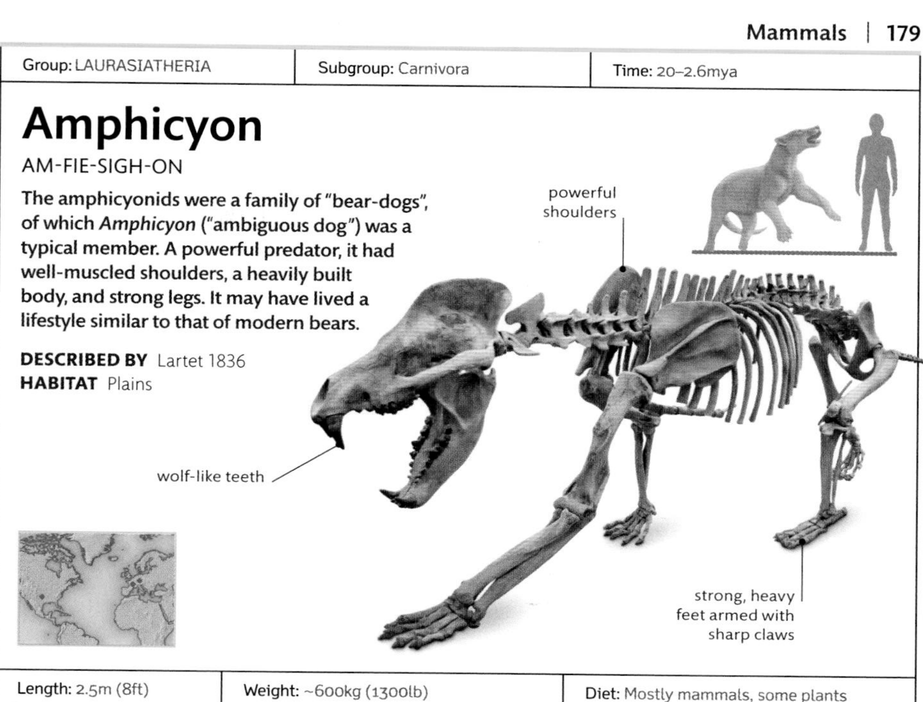

Length: 2.5m (8ft)	Weight: ~600kg (1300lb)	Diet: Mostly mammals, some plants

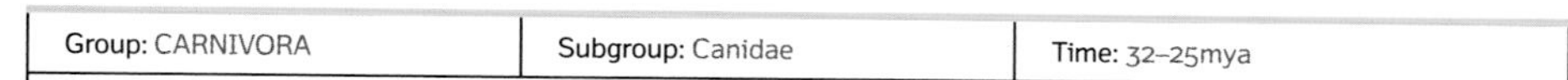

Group: CARNIVORA	Subgroup: Canidae	Time: 32–25mya

Cynodesmus

SY-NOH-DES-MUS

An early member of the dog family, *Cynodesmus* was a coyote-sized animal with a short snout and a long body. Its legs were similar to those of modern dogs, but were not as efficient for running. The feet had clawed toes, with the claws being partially retractable.

DESCRIBED BY Scott 1893
HABITAT Plains

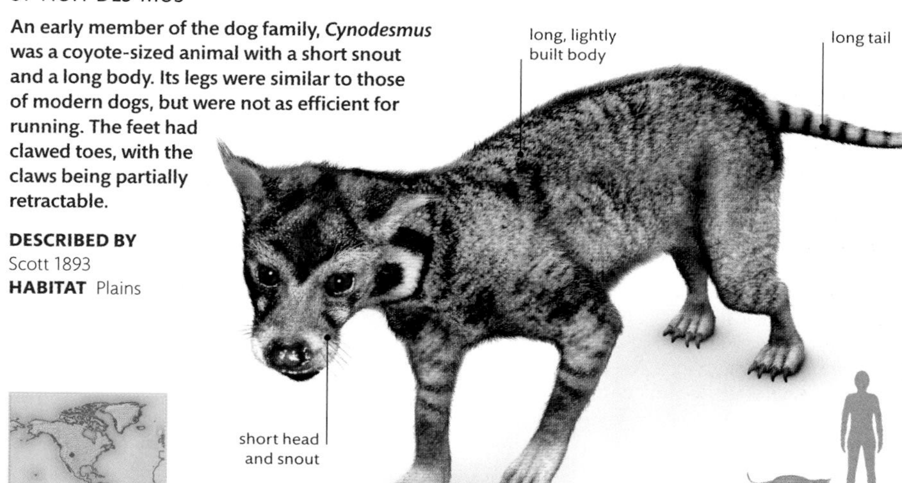

Length: 1m (3½ft)	Weight: ~70kg (155lb)	Diet: Smaller mammals, carrion

Group: LAURASIATHERIA	Subgroup: Carnivora	Time: 37–20mya

Hoplophoneus

HOP-LOE-FOE-NE-US

Hoplophoneus, a member of a family of "false sabre-tooth cats", first arose in Europe, and then gradually spread into North America. It was a large and powerful, hunting cat with a long, low body and relatively short legs. Its head and snout were short, with forward-facing eyes that probably allowed binocular vision and therefore accurate judgement of distances – an important attribute for a hunter. Its upper canine teeth were elongated into thick, stabbing "sabre" teeth, just as in the true sabre-teeth cats. These teeth were thick and curved, and extended far below the level of the lower jaw, which was capable of opening to an angle of 90 degrees, enabling it to use these teeth to stab its prey. In comparison, the lower canines were very small.

DESCRIBED BY Cope 1874
HABITAT Plains

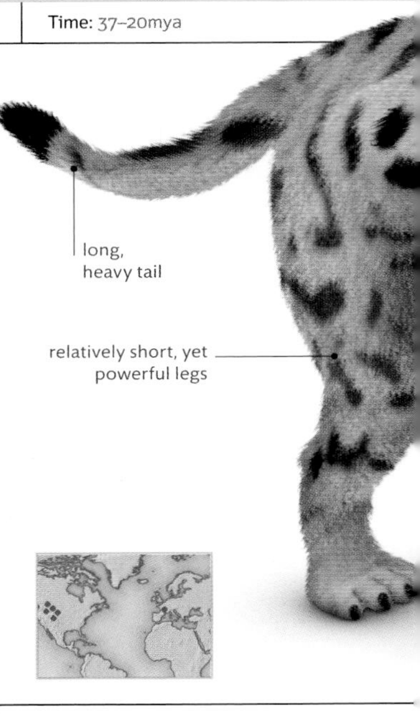

Length: 2.5m (8ft)	Weight: ~160kg (350lb)	Diet: Mammals

Group: LAURASIATHERIA	Subgroup: Perissodactyla	Time: 20–5mya

Teleoceras

TELL-EE-OH-SEA-RAS

This mammal was a member of the rhinoceros family. It had a long body, extremely short legs, and a small, conical nose horn. It is now thought that *Teleoceras* was a terrestrial grazer much like a modern rhinoceros. A large number of *Teleoceras* remains have been found in Ashfall Fossil Beds, Nebraska, USA. The animals died due to inhaling volcanic ash that buried their waterhole.

DESCRIBED BY
Hatcher 1894
HABITAT
Grasslands

low, barrel-shaped body held capacious gut

short neck

long cheek teeth

very short, but sturdy legs

Length: 4m (13ft)	Weight: ~3 tonnes (3 tons)	Diet: Shrubs, grass

Group: LAURASIATHERIA	Subgroup: Perissodactyla	Time: 16–5mya

Merychippus

MER-EE-CHIP-PUS

***Merychippus* ("ruminant horse") was the first horse to feed exclusively on grass, and the first one to have a head similar to that of modern horses. The muzzle was longer than in earlier horses, the jaw deeper, and the eyes further apart. Its neck was also longer than in earlier horses, as it spent much of the time grazing. The middle toe on each foot had developed into a hoof that did not have a pad on the bottom. In some species the outer two toes only touched the ground when running; in others, they were larger.**

DESCRIBED BY Leidy 1856
HABITAT Grassy plains

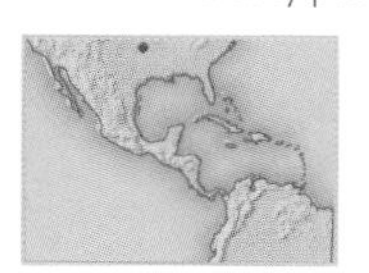

Length: 2m (6.5ft)	Weight: ~200kg (440lb)	Diet: Grass

Group: LAURASIATHERIA	Subgroup: Perissodactyla	Time: 13–2.5mya

Hipparion

HIP-PAIR-EE-ON

Hipparion ("better horse") was one of several groups of grazing three-toed primitive horses that lived in the Miocene. It was a lightly built animal that greatly resembled a modern pony, with a long jaw, and slim legs. The horse's full weight was borne on the enlarged central toe, which had developed a recognizable hoof. The two outer toes were much reduced in size and did not reach the ground. A tendon in the foot increased the springiness of the gait, giving the advantage of greater speeds. The teeth were large and high-crowned for grazing. *Hipparion* was a very successful genus of early horse, spreading over the world in large herds for millions of years.

DESCRIBED BY de Christol 1832
HABITAT Plains

Upper jaw bone

large premolars

high-crowned molars with many cusps

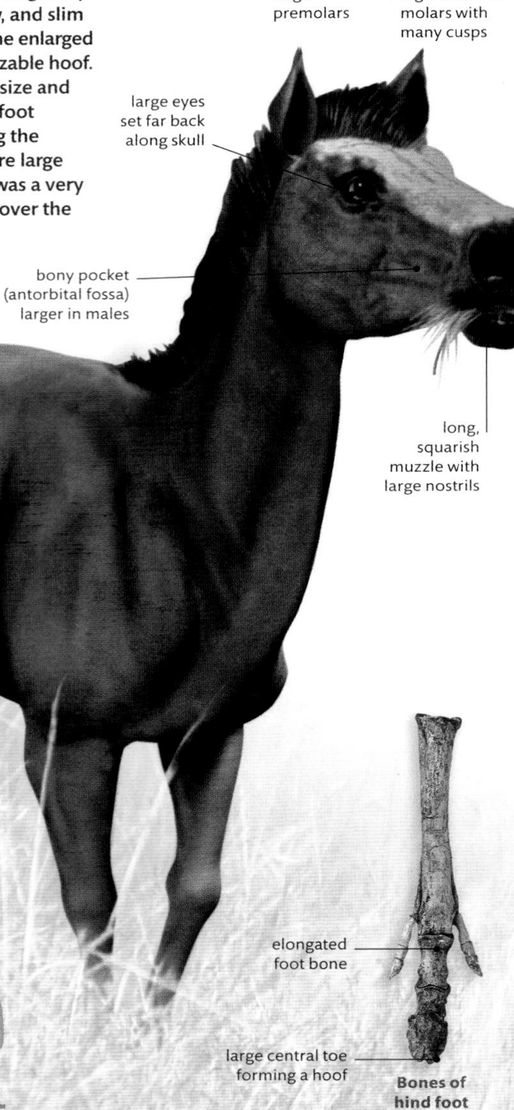

Length: 1.5m (5ft)	Weight: ~115kg (250lb)	Diet: Grass

Group: CETARTIODACTYLA	Subgroup: Cetacea	Time: 23–15mya

Eurhinodelphis

UOU-RHINE-OH-DEL-FIS

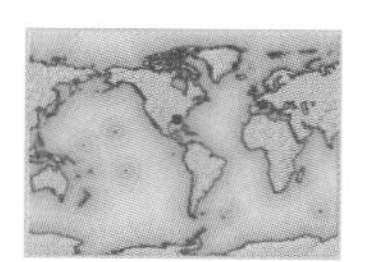

Eurhinodelphis ("true-nose dolphin") was one of the most common dolphins in the Miocene. Its most distinctive feature was its long snout, which may have been used to strike at its prey. The structure of the ear had become more complex than in earlier toothed whales, indicating that *Eurhinodelphis* had developed some form of echolocation system. The skull was slightly asymmetrical, as in modern toothed whales, with a blowhole at the top of the head.

DESCRIBED BY
Du Bus 1867
HABITAT
Oceans

asymmetrical skull

elongated snout

hydrofoil-shaped flippers

powerful, double-lobed tail

Length: 3.7m (12ft)	Weight: Not calculated	Diet: Fish

Group: CETARTIODACTYLA	Subgroup: Cetacea	Time: 14–7mya

Cetotherium

SET-OH-THEER-EE-UM

Resembling a small version of the modern rorquals, *Cetotherium* was an early baleen whale, with baleen plates instead of teeth. The inside of the baleen was edged with coarse hairs that filtered krill, plankton, and small fish. *Cetotherium's* baleen plates were probably quite short. It lacked echolocation. Its head was symmetrical and it probably had two blowholes.

DESCRIBED BY
Brandt 1843
HABITAT
Oceans

long, streamlined body like modern baleen whales

short baleen plates

elongated, wide head

Length: 4m (13ft)	Weight: ~2 tonnes (2 tons)	Diet: Plankton

BIRDS

THE SMALL SIZE and delicate skeleton of birds mean that their fossils are much less common than those of other animals. Nevertheless, several sites around the world with exceptional preservation show that a great diversity of birds had evolved by the Neogene, including a vast array of seafaring birds. Many of these, such as penguins, cormorants, and boobies, are still alive today. Others left no descendants, such as the enormous *Osteodontornis* (see p.186), which had a fearsome beak adorned with tooth-like bony projections.

Throughout the Neogene, many bird lineages became larger, most likely associated with cooling climates. This trend resulted in some of the largest birds to ever live, such as *Argentavis* (see p.186). Many flightless birds also became larger, including not only ratites such as ostriches, emus, and moa, but also ground-dwelling parrots and penguins.

By the Miocene, terrestrial bird ecosystems were structured very much like those of today. Ducks, owls, hawks, fowl, and parrots had all evolved and diversified, alongside the incredibly species-rich songbirds. By the end of the Pliocene, all of the modern orders of birds had evolved, including several genera that survive today. Some of these orders have no fossil records, but modern techniques using DNA show that they must have diverged by this time.

Phorusrhacids remained top predators in South America, and they were highly successful in the grasslands promoted by the cool, arid climates of the Miocene and Pliocene. However, they would go extinct shortly after the Great American Biotic Interchange, most likely due to competition from an influx of mammalian predators.

Group: NEOAVES	Subgroup: Phorusrhacidae	Time: 18–16mya

Phorusrhacos

FOR-US-RAKE-US

Phorusrhacos was one of the dominant land predators in South America at the time it existed. It had very strong legs, capable of running at high speed, stubby, flightless wings, a long neck, and a proportionately large head. This ended in a huge, hooked beak that could have torn through flesh easily, or have stabbed into prey. The lower jaw was smaller than the upper jaw. There were three toes on each of the feet, all of which were armed with sharp claws.

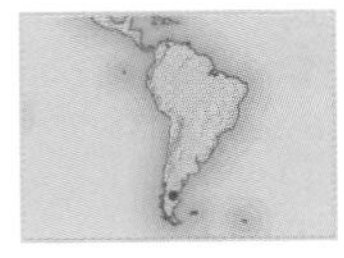

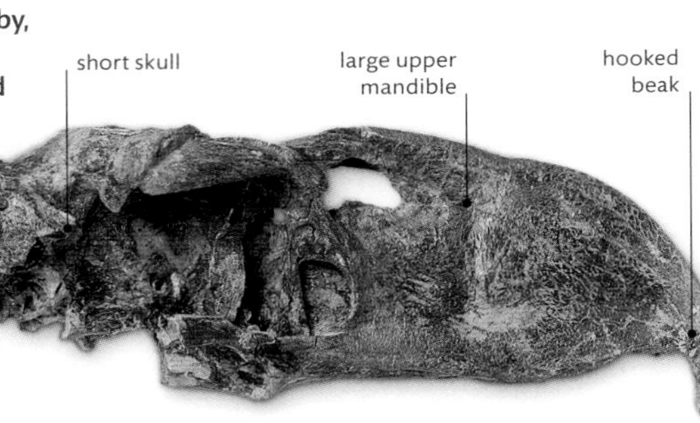

DESCRIBED BY Ameghino 1887
HABITAT Plains

Length: 1.5m (5ft)	Weight: ~80kg (176lb)	Diet: Small mammals, carrion

Group: NEOAVES	Subgroup: Phorusrhacidae	Time: 5–1.8mya

Titanis

TIE-TAN-IS

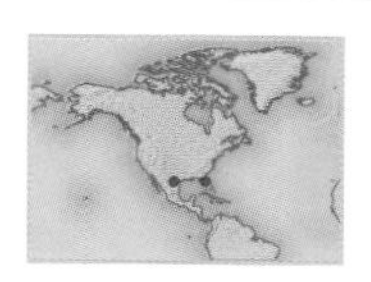

Titanis ("terror crane") was aptly named. This giant flightless bird was one of the most efficient predators of its time in North America. Its head was as large as that of a modern horse, and it had a huge, curved beak. Although it had no teeth, the sharp hook at the end of the beak was very efficient at tearing through flesh. It had highly reduced wings that were ineffective for hunting or flight. _Titanis_ had a long neck and males may have had an ornamental crest on the top of the head. It had long, agile legs, and three-toed feet with long talons. It could undoubtedly run at high speeds when hunting.

DESCRIBED BY Brodkorb 1963
HABITAT Grassy plains

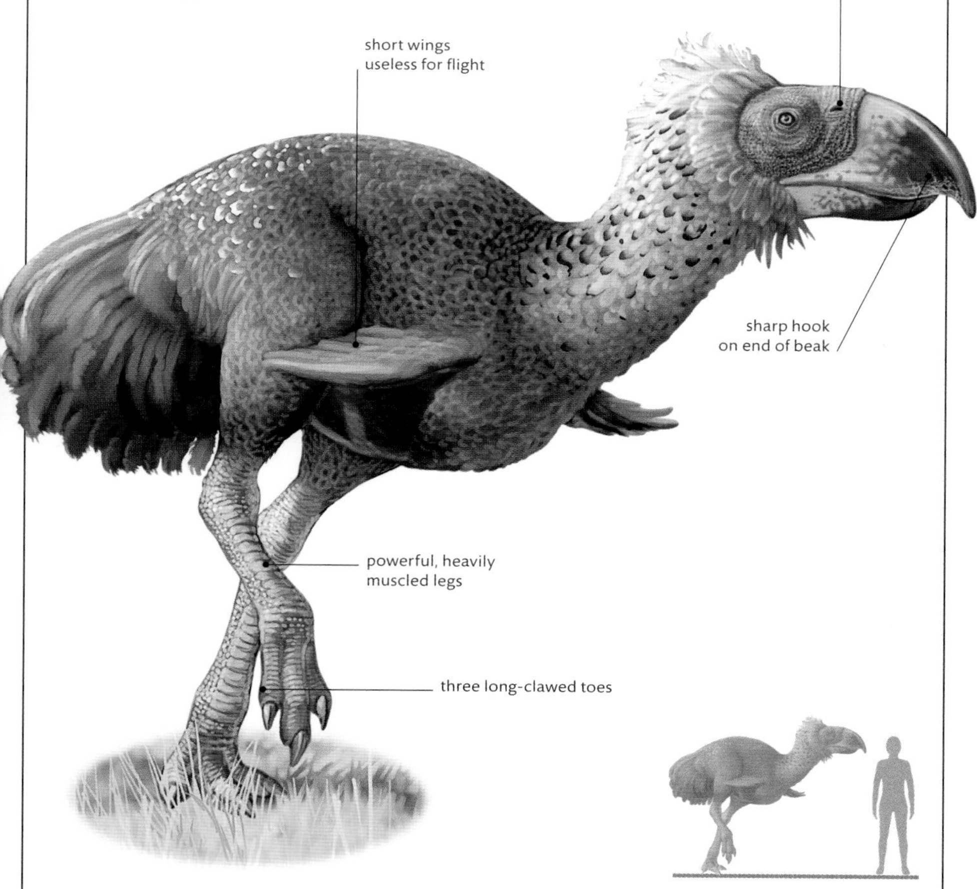

Length: 2.5m (8ft)	Weight: ~150kg (330lb)	Diet: Mammals

Group: GALLOANSERAE	Subgroup: Pelagornithidae	Time: 16–5mya

Osteodontornis

OS-TEE-OH-DON-TOR-NIS

One of the largest birds ever to have flown, *Osteodontornis* was a bony-toothed seabird. It had a large, heavy body, and long, narrow wings designed for gliding long distances. The neck had a natural S-shaped curve, meaning that the head was held over the shoulders in flight. The beak was stout and rounded, with tooth-like bony projections edging each jawbone.

DESCRIBED BY Howard 1957
HABITAT Seashores

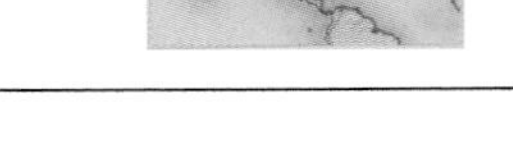

Length: 1.2m (4ft)	Weight: Not calculated	Diet: Fish

Group: NEORNITHES	Subgroup: Teratornithidae	Time: 9–6.8mya

Argentavis

AR-GEN-TAH-VIS

Only a few bones of this large bird of prey have been discovered, but these indicate a wingspan of over 7m (23ft), which is twice the size of that of the largest living bird, the wandering albatross. In appearance, it was probably much like a modern vulture, and may have been a scavenger. Its beak was large and hooked, and was used to grasp prey as its clawed feet were not well adapted for this. *Argentavis* used its vast wings to soar effortlessly, riding the warm thermals of air.

head possibly bald with neck fringe of feathers

deep, hooked beak

large eyes

DESCRIBED BY Campbell and Tonni 1980
HABITAT Inland and mountainous areas

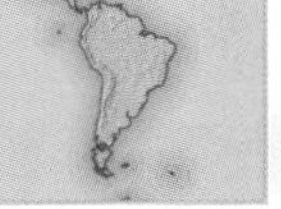

Height: 1.5m (5ft)	Weight: ~80kg (175lb)	Diet: Carrion, large herbivorous mammals

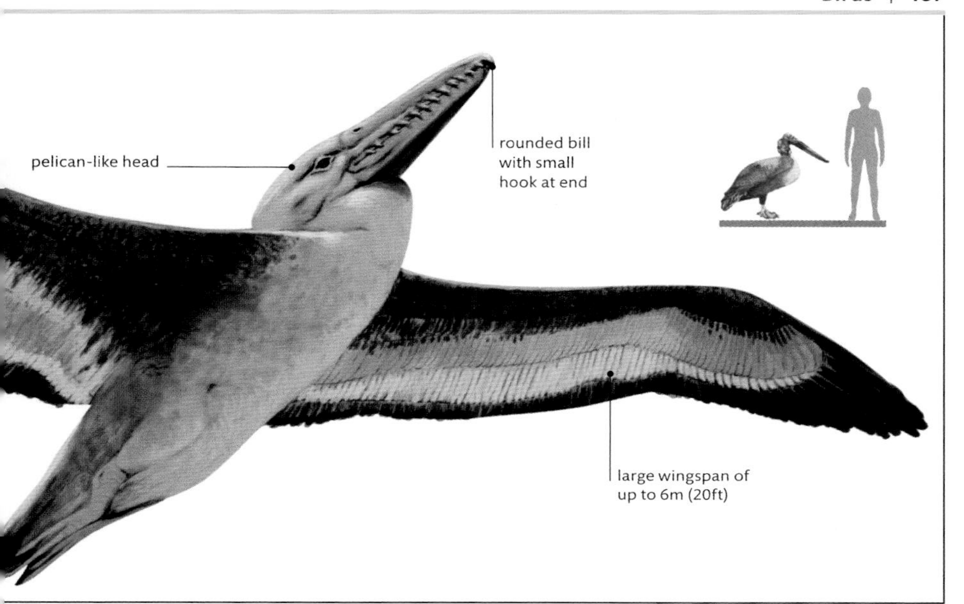
pelican-like head
rounded bill
with small
hook at end
large wingspan of
up to 6m (20ft)

wing feathers
up to 1.5m
(5ft) long
long wings compared
to body size
large feet with
three forward-
facing clawed toes

QUATERNARY PERIOD

2.6 MILLION YEARS AGO – PRESENT

THE QUATERNARY PERIOD includes the Pleistocene and Holocene Epochs. The Pleistocene was a time when the world was in the grip of a great ice age, with enormous ice sheets covering most of the northern hemisphere. Glaciers also formed in the high mountain regions of the world, including the Andes and the Himalayas. Over the next one and a half million years the Earth passed through at least four ice ages. Between ice ages, the climate was warmer. The most recent ice age ended about 10,000 years ago at the beginning of the Holocene, and saw mass extinctions of mammal species.

QUATERNARY LIFE

Animals responded to the colder conditions of the Pleistocene by migrating towards equatorial regions (for example, lizards) or evolving furry coats (for example, mammoths and the woolly rhino, *Coelodonta*). Truly modern humans, *Homo sapiens*, evolved during the Pleistocene, along with other, less successful human species, such as *Homo neanderthalensis*.

large canine teeth

powerful hindlimbs

Canine carnivores

Many species of dog lived during the Pleistocene. *Aenocyon dirus* (left), was very like a modern wolf, but heavier. It survived into the Holocene. Fossil remains show that dire wolves and sabre-tooth cats often fought over territory and food.

4,600mya 4,000mya 3,000mya

QUATERNARY LANDMASSES

During the Pleistocene Epoch, huge ice sheets covered most of North America, Europe, and northern Asia. There were land bridges from North America to Eastern Asia and between Australia and New Guinea. At the start of the Holocene, rising sea levels caused by melting ice sheets covered these land bridges. Africa and Australia continued to move northwards to occupy their present positions (see map).

Litopterns
The Pleistocene grasslands of South America were home to herds of strange, hoofed animals called litopterns. These, such as the camel-sized *Macrauchenia* (right), were grazing animals, adapted for life on open plains.

Giant marsupials
Herds of giant, grazing marsupials in Australia, such as *Diprotodon* (left), became rarer as the climate grew drier with the continent's continuing drift northwards. They survived into the Holocene, but may have been hunted to extinction by humans.

2,000mya 1,000mya 500mya 250mya 0

MAMMALS

BY THE QUATERNARY, mammal ecosystems were essentially modern. Carnivorans took over the roles of top predators, and the first big cats evolved, including *Smilodon* (see p.194). Perissodactyls had declined in diversity, possibly because of competition with ruminant artiodactyls, like deer, bovids, and giraffes, which have a more efficient digestive system. Many of these ungulates, including horses, camels, and rhinos, had ranges across the northern hemisphere. Some groups of proboscideans had gone extinct, but elephants remained successful throughout Eurasia as well as North America.

The Pleistocene ice ages had a profound impact on mammals. The spread of ice caps and glaciers forced mammal populations to move further south, and made vast swaths of land uninhabitable. Many groups of mammals became much larger and developed thick fur coats to insulate them against the freezing temperatures.

Humans continued to spread across Asia and Europe, and multiple species coexisted at the same time. The ability of these species to use tools meant that they had a profound impact on mammal populations, and there was a severe decline in the numbers of most large mammals.

At the end of the Pleistocene, there was selective extinction of many of the large-bodied mammals. Many lineages went entirely extinct, but some were able to survive in smaller parts of their ranges, resulting in the distributions we see today. It is clear that both climate change and human hunting were involved in these extinctions, as many coincided with the arrival of humans.

Group: CARNIVORA	Subgroup: Canidae	Time: 125,000–9,500ya

Aenocyon dirus

AY-NO-SY-ON DIE-RUS

This extinct member of the dog family, whose name means "dire wolf", was a larger and heavier version of the modern wolf. It had a wider head, stronger jaws, and longer teeth than modern members of the dog family. It had long, powerful legs, and its toes had blunt claws that could not be retracted. It was probably both a scavenger and an active hunter.

DESCRIBED BY (Leidy 1858)
HABITAT Grassland and woodland

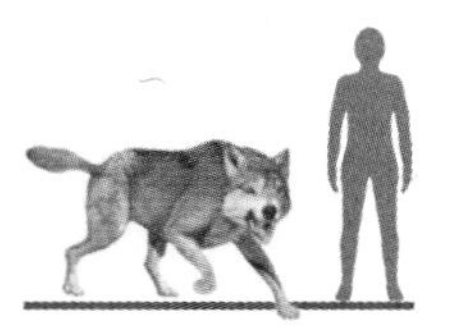

Length: 2m (6½ft)	Weight: ~53kg (115lb)	Diet: Mammals, carrion, possibly fruit

Group: AFROTHERIA	Subgroup: Proboscidea	Time: 1.5mya – 10,000ya

Mammuthus columbi

MAM-UT-US KO-LUM-BE

The largest of all mammoth species, the Columbian mammoth was probably the largest terrestrial animal alive during the last ice age. Unlike the woolly mammoth, this mammoth was well-adapted to the warmer climates it lived in and did not have such thick fur. Besides its immense size, its most distinguishing feature was its tusks. These curled backwards evenly and were up to 4.8m (15ft) long. They could weigh up to 84kg (185lb).

DESCRIBED BY (Falconer 1857)
HABITAT Plains

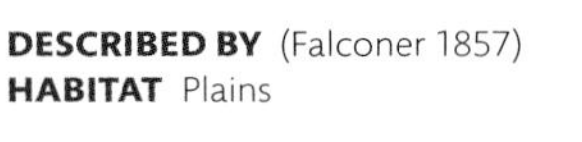

Length: 4.5m (15ft)	Weight: ~8–10 tonnes (8–10 tons)	Diet: Grass, leaves, flowering plants

Group: AFROTHERIA	Subgroup: Proboscidea	Time: 700,000–4,000ya

Mammuthus primigenius

MAM-UT-US PRIM-EE-GEN-EE-US

***Mammuthus primigenius*, the woolly mammoth, was relatively small for a mammoth. It was adapted to living in the cold climates of northern tundras, with a thick coat of long dark hair overlying a downy undercoat. The hair would have been light brown, to brown to black, depending on the individual. A thick layer of fat under the skin helped in insulation, and there was also a fatty hump behind the head used for food storage. Its ears were smaller than those of modern elephants, helping to reduce heat loss. The long, curved tusks appear to have been used to scrape away ice from the ground when feeding, as well as for protection and in dominance rituals. The woolly mammoth survived until relatively recent times, and palaeontologists are therefore very familiar with its anatomy and appearance. Several well-preserved specimens have been found frozen in the permafrost of Siberia and Alaska, and cave paintings by early humans depict it clearly.**

DESCRIBED BY (Blumenbach 1799)
HABITAT Frozen tundra

ridges for grinding tough plant material

Upper cheek tooth

long, curved tusks

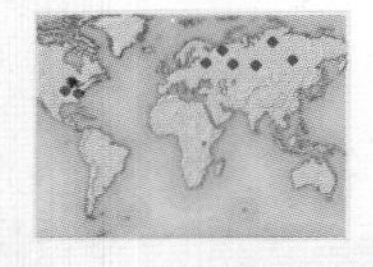

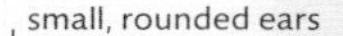

Frozen remains of a young mammoth

Length: 3.5m (11½ft)	Weight: ~6 tonnes (6.6 tons)	Diet: Low-lying tundra vegetation

Preserved hair

Group: CARNIVORA	Subgroup: Felidae	Time: 2.5mya–10,000ya

Smilodon

SMILE-OH-DON

Smilodon was the cat popularly thought of as a sabre-tooth tiger. Its sharp canine teeth were huge, and were serrated along their rear edges to increase their cutting ability. They were also oval in cross-section. This meant that they were strong while presenting minimum resistance when biting. Such long teeth meant that the cat's jaw had to open to an angle of more than 120 degrees to allow the teeth to be driven into its prey. However, it seems the teeth may have broken easily if they came into contact with bone. _Smilodon_ therefore probably hunted by biting through an animal's neck. _Smilodon's_ arms and shoulders were very powerful, in order to produce a strong downwards motion of the head. _Smilodon_ appears to have hunted in packs, probably preying on large, slow, thick-skinned animals.

DESCRIBED BY Lund 1842
HABITAT Grassy plains

FOSSIL SKULL

Smilodon is very well known. Over 2,000 individuals have been excavated from the Rancho La Brea tar pits, a famous prehistoric predator trap, in Los Angeles, California.

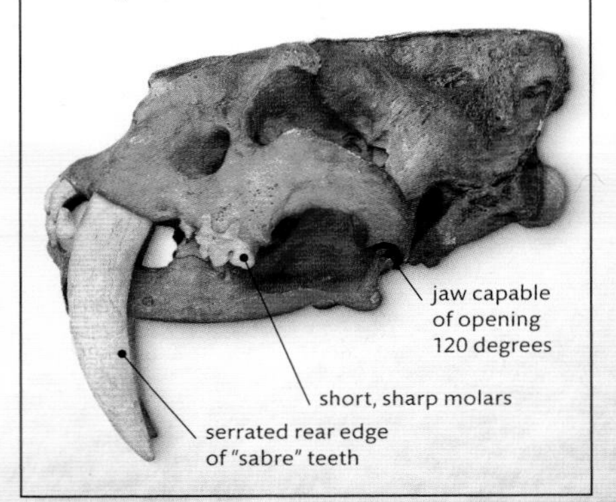

Length: 1.5–2.5m (5–8¼ft)	Weight: ~320kg (710lb)	Diet: Large mammals

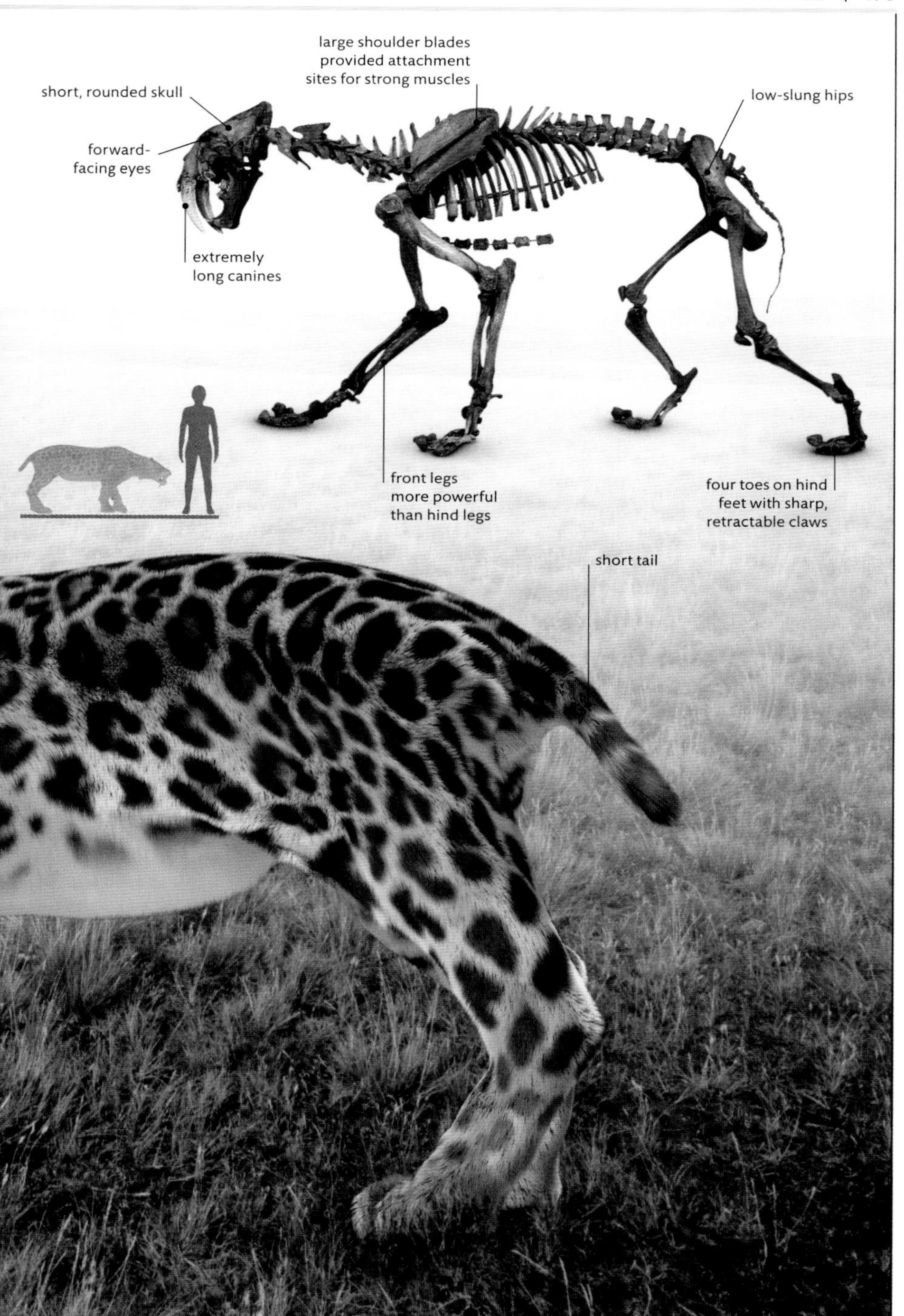
large shoulder blades provided attachment sites for strong muscles
short, rounded skull
low-slung hips
forward-facing eyes
extremely long canines
front legs more powerful than hind legs
four toes on hind feet with sharp, retractable claws
short tail

Group: CARNIVORA	Subgroup: Felidae	Time: 5–1mya

Dinofelis

DIE-NOH-FEEL-IS

Dinofelis ("terrible cat") was a panther-sized cat. It had flattened canine teeth that were intermediate in length between those of the sabre-tooth cats and those of biting cats, such as lions. It was probably an agile climber with long, strong legs, and sharp, retractable claws. Its tail was long and flexible and could be used to aid balance. The skull of *Dinofelis* was short and its eyes were forward-facing and positioned for binocular vision - essential for judging distances accurately when hunting or leaping. Its coat was likely to have a similar pattern of camouflage to those of modern forest-dwelling cats.

DESCRIBED BY
Zdansky 1924
HABITAT Dense forest

dappled coat for camouflage

flat head with forward-facing eyes

long canine teeth

retractable claws remained sharp

Length: 2.1m (7ft)	Weight: ~160kg (350lb)	Diet: Mammals

Group: CARNIVORA	Subgroup: Felidae	Time: 1mya to present

Panthera

PAN-THAIR-A

There were several lion-like species of *Panthera* that are now extinct. *Panthera spelaea*, the European cave lion, was the largest cat ever to have lived. It is often seen in cave paintings. *Panthera atrox* lived in North America. Both species killed prey by biting the neck with their canine teeth, and had sharp claws that could be retracted under a flap of skin.

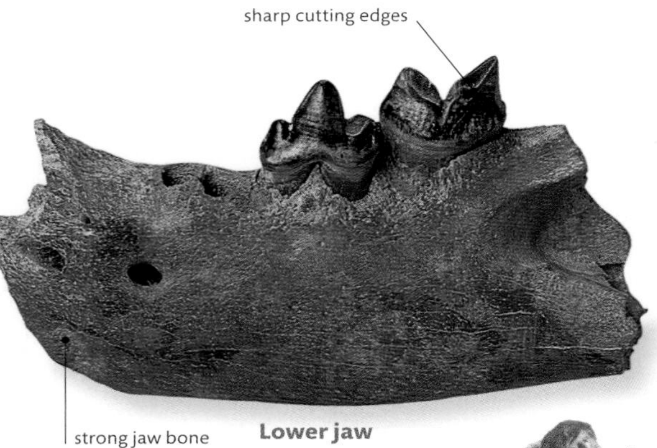

Lower jaw

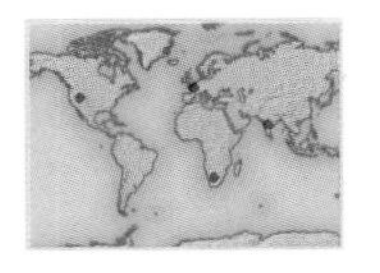

DESCRIBED BY
Oken 1816
HABITAT
Grassland

Length: 3.5m (11ft)	Weight: ~235kg (520lb)	Diet: Meat

Group: LAURASIATHERIA	Subgroup: Meridiungulata	Time: 9mya – 10,000ya

Macrauchenia

MAK-RAW-CHEN-EE-A

This curious animal (whose name means "great llama") had a number of camel-like features, such as its size, its small head, and long neck. However, its three-toed feet were more like those of a rhinoceros. Challenging the conventional interpretation of a tapir-like proboscis, it is now suggested that *Macrauchenia* had a nose similar to that of a modern moose. It had long, slim legs, and the flexibility and strength of the ankles suggest it was a relatively agile animal. The high-crowned cheek teeth indicate that *Macrauchenia* may have been a browser and a grazing animal.

camel-like body shape

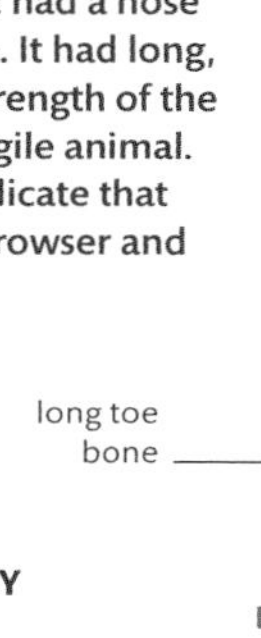

Right forefoot bones

DESCRIBED BY Owen 1838
HABITAT Woodland

Length: 3m (10ft)	Weight: ~700kg (1,540lb)	Diet: Plants

Group: LAURASIATHERIA	Subgroup: Perissodactyla	Time: 3.7mya – 10,000ya

Coelodonta

SEEL-OH-DON-TA

The woolly rhinoceros, *Coelodonta* ("hollow tooth"), evolved to withstand cold conditions. Its fur was thick and shaggy, its body was massive with short legs, and its ears were small. It had a huge pair of horns on its snout, the front one of which grew to over 1m (3¼ft) in length in older males.

thick, shaggy, grey-brown fur

small ears to reduce heat loss

long horn made of keratin

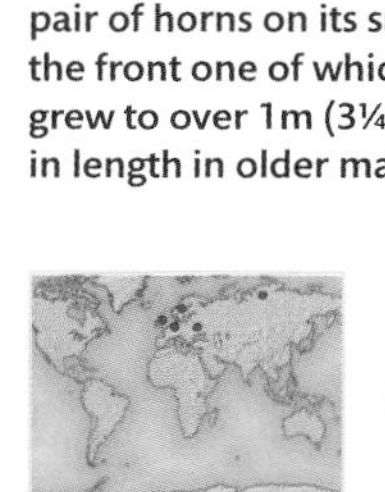

DESCRIBED BY Bronn 1831
HABITAT Tundra

Length: 3.5m (11ft)	Weight: ~3–4 tonnes (3–4 tons)	Diet: Plants

BIRDS

THE INSTABILITY OF THE climate during the transition from the Neogene to the Quaternary had a profound impact on birds. Several lineages went extinct during the glaciation of the Pleistocene, including many of the larger-bodied birds that had been successful during the Neogene. *Titanis* (see p.185), the last surviving phorusrhacid, was able to spread as far north as Texas during the Great American Biotic Interchange. However, it went extinct early in the Pleistocene, ending the reign of large, flightless avian predators.

Some large-bodied birds survived on isolated islands, including New Zealand and Madagascar, where they had no natural predators. However, as humans spread across the world and began to disperse to these islands, they found these birds to be easy prey. A great number of these species were hunted to extinction, including famous examples like the Dodo, the elephant bird, *Vorombe titan* (see p.199), and the moa, *Dinornis* (see below). Human arrival often introduced invasive predators, such as cats, dogs, and snakes, so island birds have continued to decline at alarming rates.

Habitat fragmentation caused by both glaciation and humans has disproportionately impacted birds of all sizes worldwide, but studies show that conservation efforts can help to slow their decline.

Group: NEORNITHES	Subgroup: Palaeognathae	Time: 40,000–600ya

Dinornis

DIE-NOR-NIS

This was the tallest flightless bird ever to have existed, and was one of a dozen types of moa that survived in New Zealand until modern times. It was a slow-moving bird, with a bulky body, long, heavily-built legs, and a long neck. Preserved feathers show that its plumage was mostly grey and fluffy. Its beak was short and sharp-edged, and was used to shear at vegetation. Gizzard stones were used to help digest plant material. The numbers of *Dinornis* declined rapidly after the arrival of humans in New Zealand about 10,000 years ago, and they were hunted to extinction around 600 years ago.

DESCRIBED BY (Owen 1843)
HABITAT River areas, bushland

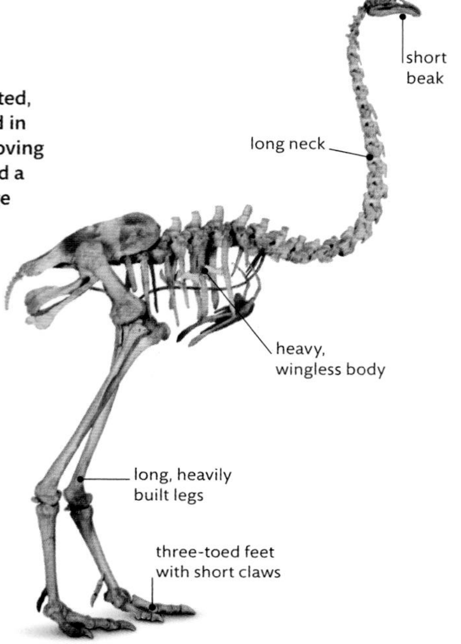

Length: 3.5m (11½ft)	Weight: ~400kg (880lb)	Diet: Twigs, seeds, and fruit

Group: NEORNITHES	Subgroup: Palaeognathae	Time: 10,000–1,000ya

Vorombe titan

VUH-ROM-BAY TIE-TAN

Vorombe titan, commonly called the "elephant bird of Madagascar" was the heaviest bird species known to have existed. It was a wingless bird, with long, thick legs and feet ending in three widely spread toes. The thickness of the thigh bone indicates that *Vorombe* was generally a slow-moving bird and could probably not run at high speeds. The small head had a toothless beak, and the only protection against predators was the bird's size and strength. Several eggs have been found fossilized in the mud of swampy areas near rivers.

DESCRIBED BY
(Andrews 1894)
HABITAT Forests

Vorombe egg

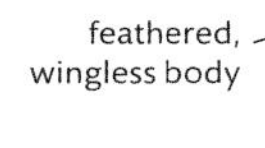

Metatarsus (foot bone)

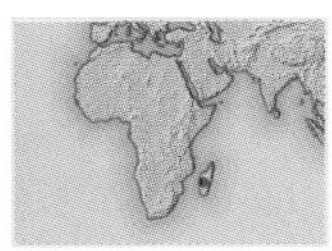

Length: up to 3m (10ft)	Weight: ~650kg (1400lb)	Diet: Seeds and fruit

ADDITIONAL DINOSAURS

This section consists of a brief description of almost 300 dinosaurs, not included in the main body of the book. The entries are arranged alphabetically and each includes, wherever possible, its geological period and the country in which the principal fossil finds were made.

Abrictosaurus
A small, early ornithopod dinosaur. Jurassic. South Africa.

Achelousaurus
A ceratopsian that appears to be an intermediate form between *Einiosaurus* and *Pachyrhinosaurus*. Late Cretaceous. USA.

Acheroraptor
A dromaeosaurid that lived alongside *Tyrannosaurus rex* and *Triceratops*. Late Cretaceous. USA.

Achillobator
A large deinonychosaur with a sickle-shaped claw on each foot. Late Cretaceous. Mongolia.

Acrocanthosaurus
An *Allosaurus*-like theropod with spines on the back supporting a sail or ridge. Early Cretaceous. USA.

Acrotholus
The earliest known pachycephalosaur. Late Cretaceous. Canada.

Adasaurus
A lightly built dromaeosaur with long, sickle-shaped claws on each hind foot. Late Cretaceous. Mongolia.

Aegyptosaurus
A sauropod known only from isolated bones. Late Cretaceous. North Africa.

Aeolosaurus
A large sauropod. Late Cretaceous. Argentina.

Afrovenator
A theropod. Cretaceous. Africa.

Agilisaurus
A lightly built, herbivorous dinosaur. Middle Jurassic. China.

Alamosaurus
The last known sauropod and North America's only known titanosaur. Late Cretaceous. USA.

Albertavenator
A troodontid dinosaur. Late Cretaceous. Canada.

Albertosaurus
A *Tyrannosaurus*-like theropod. Late Cretaceous. North America.

Alectrosaurus
A large theropod. Late Cretaceous. China and Mongolia.

Algoasaurus
A small sauropod. Early Cretaceous. South Africa.

Alocodon
A small, thyreophoran dinosaur known only from teeth. Late Jurassic. Portugal.

Altirhinus
An iguanodontid with a beak on its snout and a spiked thumb on each hand. Early Cretaceous. Mongolia.

Altispinax
A theropod with long spines along its back. Early Cretaceous. England.

Alvarezsaurus
A small, lightly built, theropod with an extremely long, thin, flat tail. Part of the Alvarezsauria, with *Shuvuuia*. Late Cretaceous. Argentina.

Alwalkeria
An early Saurischian. Late Triassic. India.

Alxasaurus
A large, primitive therizinosaur. Cretaceous. Mongolia.

Amargasaurus
A sauropod with unusually long spines running along its back. Early Cretaceous. Argentina.

Ammosaurus
A sauropod with large hands and thumb claws. Now considered the same as *Anchisaurus*. Early Jurassic. USA.

Ampelosaurus
A large, titanosaurid sauropod dinosaur. Late Cretaceous. France.

Amtosaurus
Possibly an ankylosaur, but identity and classification are doubtful. Late Cretaceous. Mongolia.

Amurosaurus
A hadrosaur similar to *Lambeosaurus*. Not yet formally described. Cretaceous. Russia.

Amygdalodon
A large sauropod. Late Jurassic. Southern Argentina.

Anasazisaurus
A hadrosaur known only from a skull. Late Cretaceous. USA.

Anchiceratops
A long-frilled ceratopsian. Late Cretaceous. Canada.

Anchiornis
A well-known, bird-like paravian. Late Jurassic. China.

Andesaurus
An enormous titanosaurid sauropod. Cretaceous. Argentina.

Anserimimus
An *Ornithomimus*-like dinosaur. Late Cretaceous. Mongolia.

Antarctosaurus
A very large, heavily built sauropod. Late Cretaceous. South America, India, and Russia.

Anzu
A large, toothless caenagnathid with a tall crest on its head. Late Cretaceous. USA.

Apatoraptor
A caenagnathid, known from a relatively complete articulated skeleton. Late Cretaceous. Canada.

Aquilops
A primitive neoceratopsian ornithischian. Early Cretaceous. USA.

Aragosaurus
A huge sauropod. Early Cretaceous. Spain.

Aralosaurus
A hadrosaur ornithopod known only from a skull. Late Cretaceous. Kazakhstan.

Archaeornithoides
A theropod with unserrated teeth. Late Cretaceous. Mongolia.

Archaeornithomimus
An *Ornithomimus*-like dinosaur with primitive clawed fingers. Late Cretaceous. North America and China.

Argyrosaurus
A large sauropod. Late Cretaceous. South America.

Arkansaurus
A theropod, perhaps an ornithomimid. Late Cretaceous. USA.

Arrhinoceratops
A horned ceratopsian with a short nose horn. Late Cretaceous. Canada.

Arstanosaurus
A hadrosaur with a flat head. Late Cretaceous. Kazakhstan.

Asfaltovenator
A primitive tetanuran that shows that Carnosauria is a natural group. Middle Jurassic. Argentina.

Asiaceratops
Primitive ceratopsian. Fragmentary remains only. Late Cretaceous. Russia.

Atlascopcosaurus
An iguanodont. Early Cretaceous. Australia.

Aublysodon
Small *Tyrannosaurus*-like theropod with smooth teeth. Late Cretaceous. North America.

Aucasaurus
An abelisaur dinosaur known from a complete skeleton. Late Cretaceous. Argentina.

Auroraceratops
A small, primitive ceratopsian. Early Cretaceous. China.

Austrosaurus
A primitive sauropod dinosaur. Cretaceous. Australia.

Avaceratops
A small ceratopsian dinosaur, possibly a juvenile. Late Cretaceous. USA.

Avimimus
An *Oviraptor*-like theropod. May have had feathered wings. Late Cretaceous. Mongolia and China.

Bactrosaurus
A duck-billed ornithopod with a flat head and high-spined vertebrae. Late Cretaceous. China and Uzbekstan.

Bagaceratops
A very small ceratopsian. Late Cretaceous. Mongolia.

Bagaraatan
A primitive theropod. Now considered a tyrannosaur. Late Cretaceous. Mongolia.

Bahariasaurus
A tetanuran theropod. Full classification uncertain. Late Cretaceous. Egypt and Niger.

Balaur
An unusual paravian theropod with two sickle claws on each foot. Late Cretaceous. Romania.

Bambiraptor
A juvenile theropod with a sickle-shaped claw on the second toe. Originally thought to be a juvenile *Velociraptor* or *Saurornitholestes*. Late Cretaceous. USA.

Barsboldia
A large duck-billed ornithopod. Late Cretaceous. Mongolia.

Beibeilong
A giant caenagnathid known only from embryos and a nest of eggs. Late Cretaceous. China.

Beipiaosaurus
A herbivorous theropod with primitive feathers. Now known to be a therizinosaur. Cretaceous. China.

Bellusaurus
A small sauropod. May be young of another genus. Jurassic. China.

Bilkanosaurus
Bulky sauropodomorph. Late Triassic. South Africa and Lesotho.

Borealopelta
An exquisitely preserved ankylosaur, complete with skin and scales. Early Cretaceous. Canada.

Borogovia
A *Troodon*-like theropod with a sickle-shaped claw on each foot. Late Cretaceous. Mongolia.

Brachyceratops
A ceratopsian with nose horn and two brow horns. Late Cretaceous. USA.

Brachylophosaurus
Primitive hadrosaurid ornithopod with a flat crest on the top of its head. Late Cretaceous. North America.

Breviceratops
A ceratopsian. Late Cretaceous. Mongolia.

Brontosaurus
Originally thought to be the same as *Apatosaurus*. It is now recognized as a valid but separate animal. Late Jurassic. USA.

Buriolestes
An early, carnivorous sauropodomorph. Late Triassic. Brazil.

Byronosaurus
A *Troodon*-like theropod. The first of its family with unserrated teeth. Late Cretaceous. Mongolia.

Caenagnathus
A lightly built caenagnathid oviraptorosaur. Late Cretaceous. North America.

Callovosaurus
An ornithopod. Mid Jurassic. England.

Camelotia
A large sauropodomorph. Late Triassic. England.

Carcharodontosaurus
A large carnosaur. Cretaceous. Africa.

Cathetosaurus
A sauropod known only from scanty fossils. May be the same as *Camarasaurus*. Late Jurassic. USA.

Cedarosaurus
A *Brachiosaurus*-like sauropod. Late Cretaceous. USA.

Cetiosauriscus
A sauropod. Jurassic. England.

Changmiania
A primitive ornithopod known from two skeletons in sleeping poses. Early Cretaceous. China.

Chaoyangosaurus
A small, bipedal ornithischian, probably a pachycephalosaur. Late Jurassic. China.

Chilantaisaurus
A large *Allosaurus*-like carnosaur. Late Cretaceous. China and Russia.

Chilesaurus
An unusual herbivorous dinosaur that is either an early-branching ornithischian or a theropod. Late Jurassic. Chile.

Chindesaurus
A lightly built theropod. One of the earliest known dinosaurs. Late Triassic. North America.

Chirostenotes
Along with *Caenagnathus*, now known to be a caenagnathid oviraptorosaur. Late Cretaceous. Canada.

Chubutisaurus
An unusual sauropod with heavily hollowed vertebrae. Late Cretaceous. Argentina.

Chungkingosaurus
A stegosaurid. Late Jurassic. China.

Claosaurus
A primitive hadrosaur. Late Cretaceous. USA.

Coloradisaurus
A massospondylid sauropodomorph. Late Triassic. Argentina.

Concavenator
An unusual carcharodontid theropod with a hump on its back. Early Cretaceous. Spain.

Conchoraptor
A small oviraptorid theropod. Late Cretaceous. Mongolia.

Corythoraptor
An ornately-crested oviraptorid theropod. Late Cretaceous. China.

Cryolophosaurus
A theropod. Early Jurassic. Antarctica.

Dacentrurus
An early stegosaurid with two rows of asymmetrical plates along its back and large spines along its tail. Late Jurassic. England, France, Portugal, and Spain.

Daspletosaurus
A heavily built tyrannosaurid theropod with brow horns. Late Cretaceous. Canada.

Datousaurus
A *Cetiosaurus*-like sauropod with a long neck and a very solid skull. Mid Jurassic. China.

Deltadromeus
A long-limbed and fast-running tetanuran theropod. Late Cretaceous. Morocco.

Dicraeosaurus
A diplodocid sauropod with a long neck and a whip-like tail. Jurassic. Tanzania.

Dilong
A small, primitive, crested tyrannosauroid with feathers. Early Cretaceous. China.

Draconyx
An *Iguanodon*-like dinosaur. Late Jurassic. Portugal.

Dracopelta
A nodosaur. Late Jurassic. Portugal.

Dromiceiomimus
An *Ornithmimus*-like dinosaur. Cretaceous. Canada.

Dryptosaurus
The first theropod to be discovered in North America. Late Cretaceous. USA.

Echinodon
A small, early ornithischian. Late Jurassic. England.

Einiosaurus
A ceratopsian. Late Cretaceous. USA.

Elaphrosaurus
A lightly built theropod. Previously classed as an ornithomimid, now considered a noasaurid, a subgroup within Abelisauria. Cretaceous. Tanzania.

Elmisaurus
A theropod known from hands and feet. Now known to be a caenagnathid oviraptorosaur, with *Caenagnathus* and *Chirostenotes*. Late Cretaceous. Mongolia.

Emausaurus
An ornithischian with skin covered with cone-shaped and flat armour plates. Jurassic. Germany.

Enigmosaurus
A large therizinosaur. Cretaceous. Mongolia.

Eoceratops
A primitive ceratopsian with a short frill and three short facial horns. Late Cretaceous. North America.

Eolambia
An early hadrosaur-like ornithopod. Cretaceous. USA.

Eotyrannus
A theropod. Cretaceous. England.

Epachthosaurus
A very large armoured sauropod. Cretaceous. Argentina.

Erectopus
An allosauroid. Cretaceous. France and Egypt.

Erlikosaurus
A large therizinosaur. Late Cretaceous. Mongolia.

Euhelopus
A large sauropod with a very long neck and bulky body. Jurassic. China.

Euronychodon
A tetanuran theropod. Cretaceous. Portugal and Uzbekistan.

Euskelosaurus
A large sauropodomorph. Early Jurassic. Lesotho, South Africa, and Zimbabwe.

Eustreptospondylus
A large tetanuran theropod with a primitive hip structure. Jurassic. England.

Falcarius
A primitive therizinosaur known from a bonebed with thousands of bones. Early Cretaceous. USA.

Garudimimus
An *Ornithomimus*-like dinosaur. Cretaceous. Mongolia.

Gasparinisaura
A very small ornithopod, probably a juvenile. Cretaceous. Argentina.

Genyodectes
A large theropod. Cretaceous. Argentina.

Gigantoraptor
The largest oviraptorosaur, rivalling *Tyrannosaurus rex* in size. Late Cretaceous. China.

Gilmoreosaurus
A primitive ornithischian hadrosaur. Cretaceous. China.

Gobihadros
An exceptionally preserved basal hadrosauroid. Late Cretaceous. Mongolia.

Gobivenator
A beautifully preserved troodontid. Late Cretaceous. Mongolia.

Gongbusaurus
An ornithischian. Jurassic. China.

Gorgosaurus
A tyrannosaurid, once thought to be *Albertosaurus*. Cretaceous. North America.

Goyocephale
A pachycephalosaur with a thick-skull, flat-head, knobs and spikes on the skull, and large teeth. Cretaceous. Mongolia.

Gryposaurus
A hadrosaurid. Cretaceous. Canada.

Guanlong
A primitive, crested tyrannosauroid. Late Jurassic. China.

Halszkaraptor
A swan-like, semiaquatic dromaeosaur. Late Cretaceous. Mongolia.

Haplocanthosaurus
A sauropod. Late Jurassic. USA.

Haplocheirus
A primitive alvarezsaur, linking them to theropods and revealing the pattern of forelimb reduction. Late Jurassic. China.

Harpymimus
A primitive ornithomimosaur. Early Cretaceous. Mongolia.

Hesperosaurus
A primitive stegosaurid with a single row of rounded plates along its back. Originally called *Hesperisaurus*. Jurassic. USA.

Histriasaurus
A large sauropod that may have had a sail along the back. Early Cretaceous. Croatia.

Homalocephale
A pachycephalosaur with a flat skull bordered by bony knobs. Late Cretaceous. Mongolia.

Hoplitosaurus
An ankylosaur. Early Cretaceous. USA.

Hulsanpes
A small theropod known from foot bones and part of the braincase. Late Cretaceous. Mongolia

Hypacrosaurus
A large hadrosaur. Cretaceous. North America.

Hypselosaurus
A small titanosaurid sauropod, whose eggs were the first dinosaur eggs discovered. Late Cretaceous. France and Spain.

Kamuysaurus
A hadrosaur known from a complete skeleton. Late Cretaceous. Japan.

Kritosaurus
A hadrosaurid. Late Cretaceous. Argentina.

Labocania
A tetanuran theropod. Late Cretaceous. Mexico.

Laplatasaurus
An armoured sauropod that had a very long neck with grooved vertebrae. Late Cretaceous. Argentina and Uruguay.

Leaellynasaura
A small hypsilophodontid with large eyes. Early Cretaceous. Australia.

Ledumahadi
A large sauropod showing the evolution of large size before columnar legs. Early Jurassic. South Africa.

Leptoceratops
A primitive ceratopsian from Cretaceous. North America.

Limusaurus
An unusual ceratosaur that became toothless over its lifetime. Late Jurassic. China.

Lophorhothon
A hadrosaurid. Cretaceous. USA.

Losillasaurus
An enormous diplodocid sauropod. Jurassic and Cretaceous. Spain.

Lycorhinus
An early ornithopod. Jurassic. South Africa.

Magyarosaurus
A small titanosaurid sauropod. Late Cretaceous. Romania.

Majungasaurus
A large theropod with a small horn over the eyes. Late Cretaceous. Madagascar.

Malawisaurus
A sauropod. Early Cretaceous. Malawi.

Marshosaurus
A large theropod. Late Jurassic. USA.

Masiakasaurus
A theropod. Now considered a noasaurid, along with *Elaphrosaurus*. Cretaceous. Madagascar.

Megaraptor
A theropod. Cretaceous. Argentina.

Mei
A small troodontid theropod known from two skeletons in sleeping poses. Early Cretaceous. China.

Melanorosaurus
A large sauropod. Late Triassic and Early Jurassic. Lesotho and South Africa.

Metriacanthosaurus
A theropod with high dorsal spines. Cretaceous. England.

Micropachycephalosaurus
A tiny ceratopsian – one of the smallest dinosaurs known. Late Cretaceous. China.

Microvenator
The earliest caenagnathid oviraptorosaur. Early Cretaceous. China and USA.

Moros
A small, primitive tyrannosauroid. Late Cretaceous. USA.

Mussaurus
A *Plateosaurus*-like sauropodomorph known from 6m- (20ft-) long adults and remains of hatchlings – the smallest dinosaur skeletons yet discovered. Triassic. Argentina.

Muttaburrasaurus
An *Iguanodon*-like ornithopod. Early Cretaceous. Australia.

Mymoorapelta
An ankylosaurid. Cretaceous. Australia.

Naashoibitosaurus
A hadrosaur known only from a skull. May be a species of *Kritosaurus*. Cretaceous. USA.

Nanosaurus
A small ornithischian. Jurassic. North America.

Nanyangosaurus
An ornithopod dinosaur. Cretaceous. China.

Nedcolbertia
A small theropod known from three partial skeletons. Probably an ornithomimid. Early Cretaceous. USA.

Nemegtosaurus
A sauropod known from a skull, vertebrae, and hindlimb bones. Late Cretaceous. Mongolia and China.

Neovenator
An *Allosaurus*-like theropod, with a puffin-like skull profile. Early Cretaceous. England.

Neuquenraptor
A theropod with a sickle claw on each foot. Late Cretaceous. Argentina.

Nigersaurus
A primitive rebbachisaurid sauropod. Early Cretaceous. Niger.

Nipponosaurus
A small hadrosaur with a bony head crest. Cretaceous. Russia.

Noasaurus
A small theropod. Cretaceous. Argentina.

Notatesseraeraptor
An early theropod similar to coelophysids and dilophosaurids. Late Triassic. Switzerland.

Nothronychus
The first therizinosaur found outside Asia. It may have been feathered. Cretaceous. USA.

Nqwebasaurus
An ornithomimid with a very large claw on the first finger. Cretaceous. South Africa.

Nyasasaurus
Probably the first dinosaur. Triassic. East Africa.

Ohmdenosaurus
A sauropod known from one complete shin bone. Early Jurassic. Germany.

Oksoko
An unusual two-fingered oviraptorid known from a group of four individuals preserved resting together. Late Cretaceous. Mongolia.

Omeisaurus
A large *Cetiosaurus*-like sauropod with a very long neck. Jurassic. China.

Opisthocoelicaudia
A large, short-tailed titanosaur, probably the same dinosaur as *Nemegtosaurus*. Cretaceous. Mongolia.

Orodromeus
An ornithopod found with unhatched eggs and young. Its young are thought to have been able to fend for themselves after hatching. Late Cretaceous. USA.

Ozraptor
An abelisauroid theropod with three-fingered hands and a stiff tail. Jurassic. Australia.

Pachyrhinosaurus
A short-frilled ceratopsian that may have had a snout horn. Cretaceous. North America.

Panoplosaurus
A nodosaurid ankylosaur with no tail club. Cretaceous. North America.

Paranthodon
A stegosaur known from a jaw bone with teeth. Cretaceous. South Africa.

Parksosaurus
An ornithopod. Cretaceous. North America.

Patagonykus
An alvarezsaur, alongside *Alvarezsaurus* and *Shuvuuia*. Late Cretaceous. Argentina.

Patagosaurus
A primitive *Cetiosaurus*-like sauropod. Mid Jurassic. Argentina.

Pawpawsaurus
A large nodosaur with armour over most of its body but no tail club. Cretaceous. USA.

Pelecanimimus
The first ornithomimosaur discovered in Europe. It had about 220 teeth, more than any other known theropod, and fossil remains show evidence of feathers. Cretaceous. Spain.

Pelorosaurus
A sauropod known from incomplete skeletons and fossilized skin impressions. Early Cretaceous. England and France.

Phyllodon
A small *Hypsilophodon*-like dinosaur. Jurassic. Portugal.

Pinacosaurus
An ankylosaur. Cretaceous. Mongolia.

Pisanosaurus
Possibly an early ornithischian known only from fragmentary fossils. Its classification is debated. Late Triassic. Argentina.

Piveteausaurus
A megalosauroid, possibly 11m (26ft) in length. Jurassic. France.

Planicoxa
An ornithopod. Cretaceous. USA.

Poekilopleuron
A *Megalosaurus*-like theropod. Jurassic. France.

Polacanthus
A primitive ankylosaur with spines. Cretaceous. England.

Prenocephale
A *Stegoceras*-like pachycephalosaur. Cretaceous. Mongolia.

Probactrosaurus
An *Iguanodon*-like ornithopod dinosaur. Early Cretaceous. China.

Proceratosaurus
A crested theropod. Now known to be an early tyrannosaur. Jurassic. England.

Prosaurolophus
A hadrosaur. Late Cretaceous. North America.

Protarchaeopteryx
A bird-like, non-flying, theropod with feathers on the arms, most of the body, and on the short tail. Early Cretaceous. China.

Protohadros
The oldest-known iguanodontian. Late Cretaceous. USA.

Pyroraptor
A *Deinonychus*-like dinosaur. May be the same as *Variraptor*. Cretaceous. France.

Qantassaurus
A kangaroo-sized ornithopod. Cretaceous. Australia.

Quaesitosaurus
Large titanosaur known only from a partial skull. Cretaceous. Mongolia.

Quilmesaurus
A medium-sized theropod. Late Cretaceous. Argentina.

Rahonavis
A dinosaur or primitive bird with sickle-shaped toe claws and a long, bony tail. Late Cretaceous. Madagascar.

Rapetosaurus
A *Titanosaurus*-like sauropod. Late Cretaceous. Madagascar.

Rebbachisaurus
A rebbachisaurid sauropod dinosaur that may have had a sail on its back. Cretaceous. Morocco and Niger.

Regaliceratops
An unusual ceratopsian with a highly ornamented, crown-like frill. Late Cretaceous. Canada.

Rhabdodon
An *Iguanodon*-like ornithopod. Late Cretaceous. Austria, France, Romania, and Spain.

Rhoetosaurus
A *Cetiosaurus*-like sauropod. Mid Jurassic. Australia.

Ricardoestesia
A small theropod. Cretaceous. North America.

Riojasaurus
A heavily built sauropodomorph. Late Triassic and Early Jurassic. Argentina.

Ruehleia
A recently discovered primitive *Plateosaurus*-like sauropodomorph. Triassic. Germany.

Saichania
An ankylosaur with a clubbed-tail, and bony spikes and knobs running along its sides. Both body armour and belly armour have been found. Cretaceous. Mongolia.

Santanaraptor
A theropod known from partial skeleton and skin impressions. Early Cretaceous. Brazil.

Sarcolestes
An early nodosaur or ankylosaur known only from a partial lower jaw. Jurassic. England.

Sauroposeidon
A *Brachiosaurus*-like sauropod. Cretaceous. USA.

Saurornitholestes
A *Velociraptor*-like theropod. Cretaceous. Canada.

Scipionyx
A theropod known from a single hatchling that includes soft tissues. Now considered a compsognathid. Cretaceous. USA.

Secernosaurus
A hadrosaur – the first to be found in South America. Cretaceous. Argentina.

Segisaurus
A goose-sized, bird-like theropod whose collar bone was similar in structure to that of true birds. Jurassic. USA.

Segnosaurus
Now known to be a therizinosaur. Cretaceous. Mongolia.

Shamosaurus
An ankylosaur. Cretaceous. Mongolia.

Shantungosaurus
The largest-known hadrosaur. Late Cretaceous. China.

Siamosaurus
A large, sail-backed, *Spinosaurus*-like theropod. Cretaceous. Thailand.

Siamotyrannus
A theropod now included within Metriacanthosauridae. Cretaceous. Thailand.

Silvisaurus
A nodosaur with a relatively long neck, and spines jutting out from the back, and perhaps the tail. Cretaceous. USA.

Sinornithosaurus
A dromaeosaurid. The fossil had traces of downy fibres on its skin. Jurassic. China.

Spiclypeus
A ceratopsian with a highly-ornamented frill. Late Cretaceous. USA.

Stenopelix
A ceratopsian, perhaps similar to *Psittacosaurus*, known from hip and leg bones. Early Cretaceous. Germany.

Stokesosaurus
A theropod that may have been the earliest tyrannosauroid. Jurassic. USA.

Struthiosaurus
An ankylosaur. Late Cretaceous. Europe.

Supersaurus
A diplodocid sauropod. One of the longest dinosaurs yet discovered. Late Jurassic. USA.

Suskityrannus
A small, primitive tyrannosauroid. Late Cretaceous. USA.

Talarurus
An ankylosaur. Cretaceous. Mongolia.

Tangvayosaurus
A titanosaurid sauropod. Cretaceous. Asia.

Tarchia
An ankylosaur with a very large tail club and large braincase. Cretaceous. Mongolia.

Tawa
An early theropod known from complete skeletons and skulls. Late Triassic. USA.

Telmatosaurus
A hadrosaur. Cretaceous. France, Romania, and Spain.

Tendaguria
A sauropod. Jurassic. Tanzania.

Tenontosaurus
A large ornithopod with relatively long arms. Cretaceous. North America.

Thanatotheristes
A tyrannosaurid closely related to *Daspletosaurus*. Late Cretaceous. Canada.

Thescelosaurus
An ornithopod dinosaur. Cretaceous. North America.

Timimus
A theropod known from leg bones. Cretaceous. Australia.

Timurlengia
A small, early tyrannosauroid with an advanced, tyrannosaurid-like brain. Late Cretaceous. Uzbekistan.

Torvosaurus
A large theropod. Late Jurassic. USA.

Tsintaosaurus
A hadrosaur with a bony crest on the head. Cretaceous. China.

Tylocephale
A small pachycephalosaur known only from incomplete skull. Cretaceous. Mongolia.

Unenlagia
A bird-like theropod. Now known to be a member of dromaeosauridae. Cretaceous. Argentina.

Utahraptor
A large dromaeosaurid. Cretaceous. USA.

Valdoraptor
A tetanuran theropod known from footbones. Early Cretaceous. England.

Variraptor
A dromaeosaurid theropod. Late Cretaceous. France.

Venenosaurus
A titanosaurid sauropod. Cretaceous. USA.

Vespersaurus
A desert-dwelling noasaurid with an unusual, reduced foot. Late Cretaceous. Brazil.

Wannanosaurus
A tiny, primitive pachycephalosaur. Cretaceous. China.

Wendiceratops
An ornate ceratopsian with a highly decorated frill and three horns on the face. Late Cretaceous. Canada.

Wuerhosaurus
A stegosaur with smaller plates than *Stegosaurus*. Early Cretaceous. Mongolia.

Xenotarsosaurus
A theropod known from a few vertebrae and hind-leg bones. Cretaceous. Argentina.

Xiaosaurus
A small ornithischian. Jurassic. China.

Xuanhuaceratops
An ornithischian. Cretaceous. China.

Yandusaurus
Ornithopod ornithischian. Jurassic. China.

Yaverlandia
An early maniraptoran. Cretaceous. England.

Yi
A small, unusual dinosaur with bat-like wings formed of skin rather than feathers. Late Jurassic. China.

Yunnanosaurus
A large sauropodomorph – the only sauropodomorph known with self-sharpening, chisel-shaped teeth. Jurassic. China.

Zephyrosaurus
A *Hypsilophodon*-like dinosaur known from a partial skull and vertebrae. Cretaceous. USA.

Zizhongosaurus
A primitive sauropod. Jurassic. China.

Zuniceratops
An early ceratopsian with brow horns. Cretaceous. USA.

GLOSSARY

This glossary is an easy-reference guide to the technical terms that are used in this book. It focuses mainly on the terms used to describe different groups of dinosaurs and other forms of prehistoric life. It also includes scientific names used to describe anatomical features. If you cannot find the term you are looking for here, check the general index as the information you require may be found elsewhere in the book. Words in bold type within the entries have their own glossary entry.

■ Acanthodians
The earliest jawed **vertebrates**. Primitive fish, also known as spiny sharks, that lived from the Ordovician to the Carboniferous.

■ Acetabulum
The hip socket.

■ Aerofoil
The curved surface of a wing that aids flight by creating an upward force.

■ Aetosaurs
A group of early Archosauromorph that superficially resembled crocodiles, with bulky bodies and leaf-shaped teeth.

■ Agnathans
"Jawless fish" – primitive **vertebrates** that flourished in Early Palaeozoic times.

■ Algae
Primitive plants and plant-like organisms.

■ Allosaurs
Large, fairly primitive **tetanuran** (stiff-tailed) **theropods**.

■ Ammonites
Extinct, predatory, marine **invertebrates** (**cephalopods**). These animals had a shell (usually spiral-coiled) containing air filled chambers; the animal lived only in the outer chamber.

■ Amniotes
Animals whose eggs contain an amnion, a membrane that surrounds the embryo. Mammals, birds, and **reptiles** are amniotes.

■ Amphibians
Vertebrates whose young live in the water (breathing through gills), but usually live on land as adults (breathing with lungs). Living amphibians include newts, salamanders, frogs, and toads.

■ Angiosperms
Flowering plants, which produce seeds enclosed in fruit (an ovary).

■ Ankylosaurs
A family of heavily armoured, plant-eating, **quadrupedal ornithischian dinosaurs** that lived from the mid-Jurassic to the Late Cretaceous Periods.

■ Arboreal
An organism that spends most of its life in trees, off the ground.

■ Archosaurs
A major group of **reptiles** that includes the **crocodilians**, **pterosaurs**, **dinosaurs**, and **aves** (birds).

■ Arthropods
A group of **invertebrates** with **exoskeletons** made of chitin, segmented bodies, and jointed limbs. Insects, spiders and scorpions, **trilobites**, and **crustaceans** are members of this group.

■ Artiodactyls
Ungulates (hoofed mammals) with an even number of toes. Pigs, camels, deer, giraffes and cattle are modern animals in this group.

■ Aves
The scientific name for birds.

■ Baltica
An ancient continent of the Palaeozoic Era.

■ Bipedal
Walking on the hindlimbs rather than on all-four legs. See also ***Quadrupedal***.

■ Bivalves
Aquatic **molluscs** enclosed by a two-part, hinged shell.

■ Brachiopods
Marine **invertebrates** with two joined valved shells.

■ Brachiosaurids
A group of huge **sauropods** with spoon-shaped teeth and long forelimbs. They lived in the Late Jurassic and Early Cretaceous Periods.

■ Braincase
The bones of the skull that contain and protect the brain.

■ Brontotheres
Also known as titanotheres. An extinct family of large, rhinoceros-like mammals.

■ Browser
An animal that eats tall foliage (leaves, trees, or shrubs).

■ Camarasaurids
A group of relatively short-necked **sauropods** of the Jurassic and Cretaceous Periods.

■ Carinates
A group of birds with deep-keeled breastbones.

■ Carnivora
A group of sharp-toothed, meat-eating mammals, including cats and dogs.

■ Carnosaurs
A group of large **theropods** that lived during the Jurassic and Cretaceous Periods.

▪ Centrosaurines
A group of rhinoceros-like **ceratopsian dinosaurs** – most had long horns on the snout and a short neck frill.

▪ Cephalopods
Molluscs with tentacles surrounding a large head. These soft-bodied **invertebrates** include squid, octopuses, cuttlefish, and **ammonites**.

▪ Cerapods
A group of **ornithischian dinosaurs** that included the **ornithopods** and the **marginocephalians**.

▪ Ceratopsians
A group of plant-eating **ornithischian dinosaurs** with beaks and bony head frills along the back of the skull.

▪ Ceratosaurs
A major group of **theropod dinosaurs** – those in which the three hip bones (**ilium**, **ischium**, and **pubis**) are fused.

▪ Cetaceans
Marine mammals with a streamlined body and flippers for limbs. The group includes whales and dolphins.

▪ Choristoderes
A group of crocodile-like **diapsid reptiles**.

▪ Clade
A grouping of animals (or other organisms) that share anatomical features derived from the same ancestor.

▪ Coelurosaurs
A group of **tetanuran theropods** that includes the **maniraptorans**, **ornithomimids**, and **tyrannosaurids**.

▪ Condylarths
A group of herbivorous mammals that arose in the Paleogene. Some had clawed feet, others had blunt hooves. This group is not natural, but is used as a descriptive grade for **ungulate**-like (hoofed) mammals whose relationships are not known.

▪ Confuciornithids
A family of early birds similar to *Confuciornis*.

▪ Creodonts
A group of meat-eating mammals with clawed feet, a small brain, large jaws, and many sharp teeth that were the dominant carnivorous mammals during the end of the Paleogene. Although this group is not natural, it is used as a descriptive grade for carnivore-like mammals whose relationships are not known.

▪ Crocodilians
A group of **archosaurs** that includes alligators, crocodiles, and gharials. They evolved during the Late Triassic Period.

▪ Crinoids
Plant-shaped **echinoderms** (sea lilies).

▪ Crustaceans
A class of **invertebrates** with a hard **exoskeleton**, jointed legs, and a bilaterally symmetrical segmented body.

▪ Cryogenian Period
A geological time period that lasted from 720 to 635 million years ago.

▪ Cycads
Primitive seed plants that dominated Jurassic habitats. They are palm-like trees, with separate male and female plants.

▪ Cynodonts
A group of herbivorous and carnivorous **synapsids** that appeared late in the Permian Period. Mammals originated within this group.

▪ Deinonychosaurs
A group of advanced **theropod dinosaurs** with a long, sharp, sickle-shaped claw on the second toe of each hind foot. This group includes **dromaeosaurs** and **troodontids**.

▪ Diapsids
A **clade** of animals, distinguished by two temporal **fenestrae**, that includes **archosaurs** and **lepidosaurs**.

▪ Dicynodonts
Pig-like, herbivorous **therapsids** with two large tusks in the upper jaw.

▪ Dimorphism
Having two forms, for example, sexual dimorphism between males and females of the same species.

▪ Dinocephalians
Late Permian **therapsids**. Some were meat-eating, others were plant-eating or omnivores.

▪ Dinosaurs
A **clade** of **reptiles**, partly distinguished by a largely to fully open **acetabulum** and erect limbs. They were wholly terrestrial. Only one major group of dinosaurs, **aves** (birds), survived after the end of the Cretaceous.

▪ Diplodocids
A family of huge **sauropods** characterized by a small head with peg-like teeth and nostrils opening at the top of the head.

▪ Docodonts
A group of primitive mammals.

▪ Dromaeosaurids
A family of small, fast **theropods**, with large, retractable, sickle-shaped toe claws and large eyes.

▪ Echinoderms
Salt-water **invertebrates** whose living members have five arms or divisions of the body (or multiples of five).

▪ Ectotherms
Animals, often called "cold blooded", whose internal temperature changes with the surrounding environment.

■ **Elasmosaurs**
A family of long-necked **plesiosaurs**.

■ **Enantiornithes**
A group of toothless birds that lived during the Cretaceous. Their shoulder blade (scapula) and coracoid (a small bone connected to the scapula) are oriented in the opposite way to that of modern birds.

■ **Endotherms**
Animals, often called "warm blooded", that maintain a relatively constant internal temperature by generating their own body heat.

■ **Euornithes**
This group encompasses all modern birds. It also includes some extinct forms that left no descendants.

■ **Eurasia**
The continent formed by the combined land masses of Europe and Asia.

■ **Evolution**
The theoretical process by which the gene pool of a population changes in response to environmental pressures, natural selection, and genetic mutations.

■ **Exoskeleton**
A tough, outer body covering made of chitin (a type of protein) or calcium carbonate.

■ **Family**
A group of related or similar organisms. A family contains one or more **genera**.

■ **Femur**
The thigh bone.

■ **Fenestra (pl. fenestrae)**
A natural, window-like hole or opening in a bone. The skull has many fenestrae.

■ **Fibula**
The smaller of the two bones of the lower leg.

■ **Fossil**
Any direct evidence of ancient life, including preserved remains, chemical traces, or behavioural traces.

■ **Furcula**
The "wishbone" in birds.

■ **Gastralia**
Thin ribs in the belly area not attached to the backbone.

■ **Gastropods**
A class of **molluscs** with a sucker-like foot and often a spiral shell.

■ **Genus (pl. genera)**
A group of related or similar organisms. A genus contains one or more **species**. A group of similar genera comprise a **family**.

■ **Gondwana**
The southern supercontinent formed after the break-up of the supercontinent **Pangaea**. It included present-day South America, Africa, India, Australia, and Antarctica.

■ **Gracile**
Small and slender. Some species have both gracile and **robust** forms, possibly representing differences between males and females.

■ **Graptolite**
Extinct, tiny marine colonial animals, with a soft body and a hard outer covering.

■ **Grazer**
An animal that eats low-lying vegetation, such as grasses.

■ **Gymnosperms**
Seed-bearing plants that do not produce flowers.

■ **Hadrosaurs**
A group of "duck-billed" Cretaceous **quadrupedal ornithopods**.

■ **Herbivore**
An animal that eats plants.

■ **Hesperornithes**
A family of early birds similar to *Hesperornis*.

■ **Homeothermic**
An animal that keeps a fairly constant body temperature.

■ **Horsetail**
A primitive, spore-bearing plant with rhizomes that was common during the Palaeozoic and Mesozoic Eras.

■ **Hypsilphodontids**
A group of small, herbivorous **ornithopod ornithischians** that were widespread in the Jurassic and Cretaceous Periods.

■ **Iapetus Ocean**
The precursor of the Atlantic Ocean, between the ancient continents of **Laurentia** and **Baltica**.

■ **Ichthyornithes**
A family of early birds similar to *Ichthyornis*.

■ **Iguanodontids**
A group of herbivorous **ornithopod ornithischians** that were widespread in the Cretaceous Period.

■ **Ilium**
One of the three (paired) bones of the pelvis.

■ **Insectivore**
An organism (animal or plant) that eats insects.

■ **Invertebrates**
Animals without a backbone.

■ **Ischium**
One of the three (paired) bones of the pelvis.

■ **Juvenile**
A young or immature individual.

■ **K-Pg Extinction**
The mass extinction that occurred at the boundary of the Cretaceous and the end of the Paleogene.

■ **Lagomorphs**
A group of mammals that

became widespread during the end of the Paleogene and includes modern rabbits and hares.

■ Lagosuchians
A group of early **archosaurs** that were probably ancestors of the **dinosaurs**.

■ Laurasia
The northern supercontinent formed after the break up of **Pangaea**.

■ Laurentia
An ancient continent of the Palaeozoic Era.

■ Lepidosaurs
The group of **reptiles** that includes snakes and lizards.

■ Lepospondyls
Small, extinct **amphibians** that resembled salamanders or snakes. They lived through the Carboniferous and Permian Periods.

■ Litopterns
A group of extinct hoofed mammals that resembled camels and horses.

■ Lobe-finned fish
A group of fish whose fins are supported on fleshy lobes. Lobe-finned fish, or **Sarcopterygians**, appeared during the Silurian Period.

■ Lycopods
A group of **lycopsid** plants, also known as club mosses. They were at their peak in the Carboniferous Period.

■ Lycopsids
Primitive **vascular plants** that evolved during the Devonian Period.

■ Maniraptorans
A group of advanced **theropods** with bird-like characteristics. It included **dromaeosaurs**, **oviraptorids**, **troodontids**, **therizinosaurs**, and **aves**.

■ Marginocephalians
A group of **ornithischian dinosaurs** with a bony frill or shelf on the back of the skull. This group includes the **ceratopsians** and **pachycephalosaurs**.

■ Marsupials
Mammals, including modern-day kangaroos, that give birth to small, undeveloped young that grow and mature in a pouch on the mother's abdomen.

■ Mastodons
An extinct group of mammals closely related to elephants.

■ Megalosaurs
A group of large **theropods**, less advanced than **allosaurs**.

■ Mesosaurs
Extinct, lizard-like aquatic **reptiles**.

■ Molluscs
A group of **invertebrates** including **gastropods** and **cephalopods**.

■ Monotremes
Primitive egg-laying mammals. The platypus and echidnas are the only living representatives of this group.

■ Mosasaurs
Large marine **reptiles** that lived during the Cretaceous Period.

■ Multituberculates
Rodent-like mammals of the Late Jurassic to Paleogene Periods. All were very small.

■ Myriapods
A group of many-legged **arthropods** that includes centipedes and millipedes.

■ Nautiloids
Primitive **cephalopods** with thick shells. Only a single **genus** survives.

■ Neognathae
A group of birds that evolved during the Late Cretaceous. They include most flying birds plus swimming and diving birds such as modern penguins.

■ Neornithes
The **clade** of modern birds (and their common ancestor) that have feathers, a beak covered in horn, and a four-chambered heart.

■ Nodosaurs
A group of **quadrupedal ornithischian** armoured **dinosaurs**.

■ Nothosaurs
Extinct marine **reptiles** with four paddle-like limbs that lived during the Triassic Period.

■ Orbit
The eye socket.

■ Ornithischians
"Bird-hipped" **dinosaurs** – one of the two main dinosaur groups. They were **herbivores** and had hoof-like claws. See also ***Saurischians***.

■ Ornithodirans
The **archosaur clade** that includes **dinosaurs**, their early ancestors the **lagosuchians**, and **pterosaurs**.

■ Ornithomimids
A group of **theropod dinosaurs** whose name means "bird-mimic". They outwardly resembled flightless birds.

■ Ornithopods
A group of **ornithischian dinosaurs** that have no hole in the outer, lower jaw and a long **pubis** that extends farther forwards than the **ilium**. They were beaked, mostly **bipedal herbivores**.

■ Ornithothoraces
A **clade** of birds that had evolved the alula – a feather that directs air over the upper surface of the wing.

■ Oviparous
A term used to describe animals that hatch from eggs.

■ **Oviraptorosaurs**
A **clade** of feathered **maniraptoran theropods**, many of which were toothless, that included *Caudipteryx*, *Avimimus*, caenagnathids, and oviraptorids.

■ **Pachycephalosaurs**
A group of **bipedal ornithischian dinosaurs** with immensely thick skulls.

■ **Palaeontology**
The branch of biology that studies the forms of life that existed in former geologic periods.

■ **Palaeontologist**
A scientist who studies **palaeontology**.

■ **Pangaea**
A supercontinent consisting of all of Earth's landmasses that formed at the end of the Palaeozoic Era.

■ **Parareptile**
A group of extinct **amniotes** considered not to be true **reptiles**.

■ **Pareiasaurs**
A group of early **parareptiles** with massive, heavy bodies and stout limbs.

■ **Permineralization**
The process in which minerals are deposited within the fabric of a bony fossil.

■ **Perissodactyls**
A group of hoofed mammals that includes horses, rhinoceroses, and tapirs.

■ **Petrification**
The process by which organic tissue turns to stone.

■ **Phytosaurs**
A group of extinct semi-aquatic **archosaurs** superficially resembling crocodiles.

■ **Placentals**
Mammals whose unborn young are nourished by an organ called a placenta.

■ **Placoderms**
A class of jawed fish protected by armour plates.

■ **Placodonts**
Aquatic **reptiles** that lived in shallow seas during the Triassic, becoming extinct at the end of the period. Many had turtle-like shells.

■ **Plesiosaurs**
Large marine **reptiles** of the Mesozoic Era with pairs of flipper-shaped limbs.

■ **Pliosaurs**
Large, short-necked **plesiosaurs**.

■ **Predator**
An animal that preys on other animals for food.

■ **Prehensile**
Able to grasp something by wrapping around it. An animal with a prehensile tail is able to grasp branches with it, for example.

■ **Primitive**
Having characteristics similar to those of earlier forms.

■ **Proboscideans**
A group of mammals with trunks that includes modern elephants as well as the now-extinct mammoths.

■ **Procolophonids**
A group of early herbivorous **parareptiles** of the Triassic Period.

■ **Protozoans**
General term for single-celled animals.

■ **Psittacosaurs**
A group of **bipedal**, herbivorous **ceratopsians** with a parrot-like beak.

■ **Pterodactyloids**
Short-tailed **pterosaurs** that replaced earlier long-tailed forms.

■ **Pterosaurs**
Flying **archosaurs** closely related to the **dinosaurs**.

■ **Pubis**
A bone that is part of the pelvic girdle. It points downwards and slightly towards the front in **saurischians** and downwards and towards the tail in **ornithischians**.

■ **Pygostyle**
The short tailbone of a bird, formed from fused tail **vertebrae**.

■ **Quadrupedal**
Walking on four legs. See also ***Bipedal***.

■ **Radius**
One of the two lower arm bones.

■ **Ratites**
A group that includes most large flightless terrestrial birds, such as ostriches, emus, and their kin.

■ **Reptiles**
The common name for a group of animals characterized by having scales (or modified scales), being ectothermic, and laying eggs with shells. This is not a natural group as it does not include all of its descendants. See also ***Ectotherms***.

■ **Rhynchosaurs**
Herbivorous, land-dwelling **diapsid reptiles** from the Late Triassic Period.

■ **Robust**
Large form of a **species**. See also ***Gracile***.

■ **Sacrum**
The **vertebrae** of the lower back fused to the pelvis.

■ **Sarcopterygians**
A group of bony fish with fleshy fins, including lungfish, coelacanths, and many extinct forms. See also ***Lobe-finned fish***.

■ **Saurischians**
"Lizard-hipped" dinosaurs – one of the two main **dinosaur** groups. Saurischians are divided into the **theropods** and **sauropodomorphs**. See also ***Ornithischians***.

■ **Sauropodomorphs**
Large, four-legged, long-necked herbivorous **saurischian dinosaurs**. They gave rise to the enormous **sauropods**.

■ **Sauropods**
Huge, **quadrupedal**, herbivorous **dinosaurs** with long necks, small heads, and long tails.

■ **Sauropterygians**
An group of extinct aquatic **reptiles** from the Mesozoic Era. They include **plesiosaurs**, **nothosaurs**, and **placodonts**.

■ **Sclerotic ring**
A ring of bones that supports the structure of the eye.

■ **Scute**
A bony plate with a horny outer covering embedded in the skin.

■ **Seed ferns**
Primitive seed plants (Pteridosperms) that grew in swampy areas through the Palaeozoic and Mesozoic Eras.

■ **Sirenians**
A group of mammals that includes modern sea cows (manatees).

■ **Species**
In Linnaean classification, the level below a **genus**. Only individuals within the same species can breed and produce viable, fertile young.

■ **Stegosaurs**
Four-legged **ornithischian dinosaurs** with bony plates and/or spikes down the neck, back, and tail.

■ **Sternum**
The breastbone.

■ **Synapsids**
A group of **tetrapod vertebrates** distinguished by having a skull with a low opening behind the eyes. Synapsids include mammals and many other Permian and Triassic forms.

■ **Tarsal**
Ankle bone.

■ **Teleosts**
A group of advanced bony fish.

■ **Temnospondyls**
A group of early **tetrapods**.

■ **Tetanurans**
One of the major groups of **theropod dinosaurs**. The rear part of the tails of tetanurans were stiffened by interlocking bony ligaments on the **vertebrae**.

■ **Tethys Sea**
A shallow sea that existed during the early Mesozoic Era, separating the northern landmass of **Laurasia** from **Gondwana** in the south.

■ **Tetrapods**
Four-legged **vertebrates** and the vertebrates descended from them.

■ **Thalattosaurs**
Large, lizard-like marine **reptiles** that lived in the Triassic Period.

■ **Therapsids**
A general term used to describe **synapsids** in the Permian and Triassic Periods. This group includes mammals.

■ **Therizinosaurs**
A group of strange **theropod dinosaurs** that had a toothless beak and four toes on each foot.

■ **Theropods**
A suborder of **saurischian dinosaurs**. They were **bipedal** and mostly carnivores.

■ **Thyreophorans**
Armoured, plated, and/or spiked **ornithischian dinosaurs**. They included the **ankylosaurs** and **stegosaurs**.

■ **Tibia**
The shin bone.

■ **Triconodonts**
An extinct group of small, early mammals that lived from the Triassic until the Cretaceous Period.

■ **Trilobites**
Early **arthropods** with external skeletons divided into three lobes.

■ **Troodontids**
A group of small, lightly built, long-legged **maniraptoran theropods** with unusually large **braincases**.

■ **Tyrannosaurids**
A family of very large **coelurosaur theropods** with two-fingered hands, small arms, a large head, sharp teeth, and powerful hind legs.

■ **Ulna**
One of the two bones of the lower arm.

■ **Ungulates**
Hoofed mammals such as horses.

■ **Vascular plants**
Land-living plants with a specialized system of tubes that carry water and nutrients.

■ **Vertebrae (sing. vertebra)**
The linked bones that form the backbone in **vertebrates**.

■ **Vertebrates**
Animals with a backbone and a skull made of bone or cartilage.

INDEX

A B

C

D

I J K

L

M

N O

U V W

X Y Z

ACKNOWLEDGMENTS

Dorling Kindersley would like to thank Alison Woodhouse for proofreading the text and Jane Parker for compiling the index.

REVISED EDITION
For their help in preparing the revised edition Dorling Kindersley would like to thank:
Sue Butterworth for compiling the index; Anjali Sachar, Arshti Narang, and Debjyoti Mukherjee for design assistance; Ankita Gupta, Chhavi Nagpal, and Tina Jindal for editorial assistance; Deepak Negi for picture research assistance; and Suhita Dharamjit (Senior Jacket Designer), Rakesh Kumar (DTP Designer), Priyanka Sharma (Jackets Editorial Coordinator), and Saloni Singh (Managing Jackets Editor) for help with the jacket.

PICTURE CREDITS
The publisher would like to thank the following for their kind permission to reproduce their photographs:
(Abbreviations key: b=bottom, c=centre, l=left, r=right, t=top, b/g=background)

5: 123RF.com: Mark Turner (tr). Getty Images / iStock: Vac1 (br); **10**: Queensland Museum; **11**: Royal Tyrrell Museum, Canada (br, cr); **17:** Alamy Stock Photo: Sergey Krasovskiy / Stocktrek Images (fcra). Nobumichi Tamura / Stocktrek Images (cra). Dorling Kindersley: James Kuether (cr); **22/23**: Corbis /ML Sinibaldi; **24/25**: Corbis/Yann Arthus-Bertrand (b/g); **27**: Natural History Museum (cl); **29:** Dorling Kindersley: Natural History Museum, London (clb); **30/31**: Corbis/Michael & Patricia Fogden (b/g); **30**: Natural History Museum (cl); **31:** 123RF.com: Corey A Ford (crb). Dorling Kindersley: James Kuether (br); **32/33**: Getty Images/William J. Hebert (b/g); **34/35**: Corbis/Michael & Patricia Fogden (b, c). Getty Images/Harvey Lloyd (t); **36/37**: Getty Images/Harvey Lloyd (b/g); **36:** Getty Images / iStock: breckeni (br); **37**: Natural History Museum, London (br); **38:** Masato Hattori; **42:** Getty Images / iStock: Vac1 (cr, b); **43:** Dorling Kindersley: Institute of Geology and Palaeontology, Tubingen, Germany (b). Corbis (tr). Natural History Museum (tc); **48:** © cisiopurple / cisiopurple.deviantart.com: (t); **49:** Dorling Kindersley: James Kuether (cr). Getty Images / iStock: breckeni (b); **53**: Carnegie Museum of Art, Pittsburgh (cr); **55**: American Museum of Natural History (t). Yorkshire Museum (b); **57:** Masato Hattori (b). Natural History Museum, London (cr); **58:** Science Photo Library: Mark P. Witton (br); **58/59**: Corbis:/Michael & Patricia Fogden (b/g); **61:** Dorling Kindersley: James Kuether (tr); **62**: Science Photo Library (b); **64/65:** Dreamstime.com: Jeroen Bader (beach); **64**: Smithsonian Institution (br);-**65**: State Museum of Nature (tc); **67:** 123RF.com: Corey A Ford (cb). Carnegie Museum of Art, Pittsburgh (br); **68/69**: 123RF.com: Tommaso Lizzul (b/g). Dorling Kindersley: James Kuether; **71:** Dorling Kindersley: James Kuether (cb). **76**: Carnegie Museum Of Art, Pittsburgh (cr); **77**: The Institute of Archaeology, Beijing (cr); **79**: American Museum of Natural History (t); **82:** Dorling Kindersley: James Kuether (b); **86/87:** Dreamstime.com: Jeffrey Holcombe (b/g); **87**: Royal Tyrrell Museum, Canada (b). Senekenberg Nature Museum (t); **88**: Royal Tyrrell Museum, Canada (c); **90/91** Dorling Kindersley: James Kuether (b); **97**: Hunterian Museum **(tr).** Natural History Museum, London (b, br); **99**: Natural History Museum, London (br); **101**: Science Photo Library: Mark P. Witton (t). Natural History Museum, London (br); **103**: Natural History Museum, London (b); **104/105**: Corbis (b/g); **105:** Alamy Stock Photo: Nobumichi Tamura / Stocktrek Images (cr); **107:** 123RF.com: Corey A Ford (br); **108:** © cisiopurple / cisiopurple. deviantart.com: (b); **110/111:** Dreamstime.com: Omdeaetb (b/g); **111**: Natural History Museum, London (c); **112/113**: Dreamstime. com: Pniesen (b/g); **113**: Senekenberg Nature Museum (tr); **114:** Dorling Kindersley: James Kuether (cra); **115:** Dorling Kindersley: James Kuether (br); **118:** Dorling Kindersley: James Kuether (br). American Museum of Natural History (tr); **119:** Dorling Kindersley: James Kuether; **120:** Dorling Kindersley: James Kuether (cr); **121:** Alamy Stock Photo: Emily Willoughby / Stocktrek Images (ca). Dorling Kindersley: Royal Tyrrell Museum of Palaeontology, Alberta, Canada (b). Alamy Stock Photo: Emily Willoughby / Stocktrek Images (ca). © cisiopurple / cisiopurple.deviantart.com: (br). Dorling Kindersley: Royal Tyrrell Museum of Palaeontology, Alberta, Canada (b); **123:** © cisiopurple / cisiopurple.deviantart.com: (cra); **127:** 123RF.com: Michael Rosskothen (cra); **129**: Queensland Museum (t). Royal Tyrrell Museum, Canada (br, b); **131**: Natural History Museum, London (b). Witmer Laboratories (tr). Natural History Museum, London (cr); **133**: Royal Tyrrell Museum, Canada (c); **134**: Royal Tyrrell Museum, Canada (tr); **136**: Dorling Kindersley: James Kuether (br). Natural History Museum, London (cr); **138/139**: Getty Images (b/g); **138**: Royal Tyrrell Museum, Canada (tr); **139**: Natural History Museum, London (tr); **140**: American Museum of Natural History (tr); **141**: Natural History Museum, London (c). Science Photo Library (b); **142/143:** Dorling Kindersley: James Kuether. Dreamstime.com: Günter Albers (b/g); **143**: Natural History Museum, London (tr, cr); **144/145:** Dorling Kindersley: James Kuether (t); **145**: Photo of CRSL. Ligabue (tr); **147**: Smithsonian Institution (cr); **150/151**: Getty Images (b/g); **152/153:** Getty Images: Sergey Krasovskiy / Stocktrek Images (t); **152** Getty Images: Sergey Krasovskiy / Stocktrek Images (cl); **154:** 123RF.com: Mark Turner (cr); **157:** Alamy Stock Photo: Nobumichi Tamura / Stocktrek Images (ca); **158/159**: Corbis (b/g); **160/161**: Corbis (b/g); **163**: Natural History Museum, London (t); **164**: American Museum of Natural History (b); **167**: Natural History Museum, London (tr); **169**: Natural History Museum, London (tr); **170/171** Dorling Kindersley: James Kuether (bc); **172**: Natural History Museum, London (c); **174/175**: Corbis (b/g); **174** Alamy Stock Photo: Sergey Krasovskiy / Stocktrek Images (bl); **177**: Willem van der Merwe: (cra); **178**: Corbis (b); **179:** Science Photo Library: Mauricio Anton (tr). Clemens v. Vogelsang: (cra); **180**: American Museum of Natural History (b); **184**: Natural History Museum, London **(b)**; **185** Alamy Stock Photo: Sergey Krasovskiy / Stocktrek Images (b); **186** Science Photo Library: Jaime Chirinos (t). **188/189**: Corbis (b/g); **192**: Natural History Museum, London (tr); **192/193**: Dreamstime.com: Mathiasrhode (b/g); **194/195**: Corbis (b/g); **198:** Alamy Stock Photo: Natural History Museum, London (br); **199**: Natural History Museum, London (br).